CYBERSECURITY AND INFORMATION SECURITY ANALYSTS

PRACTICAL CAREER GUIDES

Series Editor: Kezia Endsley

CYBERSECURITY AND INFORMATION SECURITY ANALYSTS

A Practical Career Guide

KEZIA ENDSLEY

ROWMAN & LITTLEFIELD
Lanham • Boulder • New York • London

Published by Rowman & Littlefield
An imprint of The Rowman & Littlefield Publishing Group, Inc.
4501 Forbes Boulevard, Suite 200, Lanham, Maryland 20706
www.rowman.com

6 Tinworth Street, London, SE11 5AL, United Kingdom

British Library Cataloguing in Publication Information Available

Library of Congress Cataloging-in-Publication Data

Names: Endsley, Kezia, 1968– author.
Title: Cybersecurity and information security analysts : a practical career guide / Kezia
  Endsley.
Description: Lanham : Rowman & Littlefield, [2021] | Series: Practical career guides
  | Includes bibliographical references. | Summary: "This book explores the growing,
  lucrative, exciting field of cybersecurity (also called information security or InfoSec) in
  an approachable, interesting, inviting way, geared towards middle school and high school
  students."—Provided by publisher.
Identifiers: LCCN 2020031083 (print) | LCCN 2020031084 (ebook) | ISBN
  9781538145128 (paperback) | ISBN 9781538145135 (epub)
Subjects: LCSH: Computer networks—Security measures—Vocational guidance. | Data
  protection—Vocational guidance. | Computer crimes—Prevention—Vocational
  guidance. | Computer science—Vocational guidance.
Classification: LCC TK5105.59 .E54 2021 (print) | LCC TK5105.59 (ebook) | DDC
  005.8—dc23
LC record available at https://lccn.loc.gov/2020031083
LC ebook record available at https://lccn.loc.gov/2020031084

♾™ The paper used in this publication meets the minimum requirements of American
National Standard for Information Sciences—Permanence of Paper for Printed Library
Materials, ANSI/NISO Z39.48-1992.

Contents

Introduction

Welcome to Cybersecurity and Information Security

Welcome to a career in cybersecurity, also known as information security (and often abbreviated to infosec). If you are interested in a career in this challenging, exciting, rewarding, and exponentially growing field, you've come to the right book! This book is an ideal start for understanding the various careers available to you within the infosec umbrella. It discusses the paths you should consider following to ensure you have all the training, education, and experience needed to succeed in your future career goals.

Information security is a burgeoning career that has roots in all aspects of life and business. *Getty Images/metamorwork*

At its core, information security can be described very simply as the process of "preventing unauthorized access, use, disclosure, disruption, modification, inspection, recording, or destruction of *information*," where *information* can be physical or electronic.[1]

The responsibilities of the infosec personnel at any organization include, at the least, establishing business processes that will protect information and company assets from intruders, monitoring and reacting to attacks, and maintaining and updating security standards to meet new and ongoing threats.

A Career in Infosec

A career in infosec is almost impossible to describe in just a few sentences, because it can contain such a wide variety of roles, responsibilities, and locations. Here are just a few characteristics of a career in infosec:

- You will work with many different kinds of people—people of all ages, experience levels, and roles within the company. Many of them might find the work you do securing information and assets to be a hindrance to them getting their jobs done. They might not realize the importance of securing the information/network/cloud and so forth, so part of your job will be to educate fellow coworkers and/or clients.
- Often, you will work on a team of other IT professionals whose skills and talents complement each other. Occasionally, however, you may work on projects by yourself.
- Sometimes your "customers"—that is, the people for whom you perform tasks and complete projects—will be people or companies who buy things from your company. Other times, your customers will be people who work for the same company that you do.
- Even though you may be out of college by the time you get your first job, your "education" will never be finished. IT professionals in general have to keep educating themselves on the latest technologies all throughout their careers. That is especially true for infosec personnel, given the nature of how quickly security threats change and grow.

A career in infosec will expose you to a fast-paced environment that requires problem-solving, expo-sure to all different kinds of people, and a never-ending education. *Getty Images/SARINYAPINNGAM*

Because infosec is such a broad career path, there are many different ways to divide it into categories. For simplicity, this book breaks infosec into five main areas. (But keep in mind that you can ask five different people who do this for a living how to break this career into categories and you'll get five slightly different answers.)

- *Security analysts/engineers:* These specialists analyze and evaluate weaknesses in the infrastructure (might be the software, hardware, networks, or cloud). They find the best tools and countermeasures to address those vulnerabilities. They may also assess the damage from security incidents and recommend solutions and best practices. They may also test for compliance with security policies and help the organization create, implement, or manage its security solutions.
- *Security architects:* They design the security system or major components of the security system, and they may head a security design team that's building a new security system.

- *Security administrators:* They install and manage organization-wide security systems. They may also do some of the tasks of a security analyst in smaller organizations.
- *Security software developers:* They develop security software, including tools for monitoring, traffic analysis, intrusion detection, virus/spyware/malware detection, antivirus software, and so on. They may also integrate/implement security into applications and software.
- *Cryptographers/cryptologists/cryptanalysts:* These professionals use encryption to secure information or build security software, develop stronger encryption algorithms, and/or analyze the encrypted information to break the code/cipher or to determine the purpose of malicious software. Sometimes they are also called *digital forensics experts.*[2]

You may have noticed that the duties of several of these titles overlap somewhat. That's because the use of titles and responsibilities is very fluid across boundaries, depending on many factors, including the size of the organization, how sensitive the data is, how exposed the network and company information is in general, the structure of the company's assets, the countries in which the company operates, and many other factors. In addition to the five main job titles described in the book, the textbox titled "A Rose by Any Other Name" lists many, many job titles you might see within infosec that are interchangeable with the titles used in this book. Note that as the infosec domain expands and develops further, new roles and titles are likely to emerge, and the roles attributed to the current titles will likely mature and evolve.

A ROSE BY ANY OTHER NAME

Jobs in infosec go by various titles, depending on the industry, the size of the organization, the type of data needing security, and many other factors. Here's a list of job titles you may see that all tie back to the security of information.

- Assurance validator
- Chief information security officer (CISO)
- Computer security incident responder
- Cryptanalyst

- Cryptologist
- Cybersecurity analyst
- Data security analyst
- Digital forensics expert
- Disaster recovery specialist
- Ethical hacker
- Information assurance analyst
- Information security analyst
- Internet security analyst
- Intrusion detection specialist
- IT security consultant
- IT security engineer
- Network security analyst
- Penetration tester
- Security administrator
- Security analyst
- Security architect
- Security auditor
- Security consultant
- Security software developer
- Security specialist
- Security systems administrator
- Source code auditor
- System security engineer
- Virus technician
- Vulnerability assessor
- White hat hacker

Note that when you see the title "security consultant/specialist," this more general job title can refer to any one or all of these other roles/titles. These folks are tasked with protecting computers, networks, software, data, and/or information systems against viruses, worms, spyware, malware, intrusion detection, unauthorized access, denial-of-service attacks, and so on. Regardless of the specific job area, this book uses the job title "infosec" to generally refer to this career path.

So, what are jobs in these areas like? Are jobs in one category only, or is there some overlap? What education, skills, and certifications do you need to succeed in these fields? What are the salary and job outlooks for each category? And what are the pros and cons of each type of infosec job? This book answers these questions, and many more, in the following chapters.

"The best thing about a career in infosec is that there are so many opportunities! There are so many, many jobs and different types of jobs and you can specialize in all sorts of areas—or you can be a generalist! There's great career/job security. In fact, there are currently many open jobs without a qualified person to do them." —Tanya Janca, founder, security trainer, and coach of SheHacksPurple.dev, specializing in training others in software and cloud security

The Infosec Market Today

The infosec market in the United States is in excellent shape, and it will probably remain one of the most stable, productive forces in the US job market for years to come. Between now and 2026, the US Bureau of Labor Statistics expects the infosec job market to grow 32 percent, which is much greater growth than average.[3] In addition, the median salary is also over twice as high as the average income for all jobs.[4]

The *Occupational Outlook Handbook* on the Bureau of Labor Statistics website (https://www.bls.gov/ooh) has current US information about the infosec profession. We will discuss this in further detail in chapter 1.

What Does This Book Cover?

This book takes you through the steps to see if a career in infosec may be right for you. It also gives you practical advice on how to pursue an education that will set you up to be a successful candidate for the type of job you might want.

- Chapter 1 describes the many specific paths that an infosec career can take. From network to cloud, back-end to front-end, this chapter gives you an idea of the many different types of career options that exist.
- Chapter 2 describes the education requirements that you should know as you think about entering the infosec field. It talks about steps that you can take as early as high school to prepare yourself; it also describes the things you can do outside of class to help yourself be ready for an infosec education.
- Chapter 3 looks at educational options that will lead you to a job in infosec. It discusses academic requirements, costs, and financial aid options.
- Chapter 4 helps you build the tools you need to prepare for interviewing for jobs and internships. It also helps with cover letters, explains how to dress for meetings, and helps you understand what employers expect out of people looking for jobs.

Throughout each of the chapters, you'll read interviews with real people, at various stages of their careers, who chose infosec as a career path. They offer real advice, encouragement, and ways to determine if this is something that may be right for you.

Where Do You Start?

Take a breath and jump right into the next chapter. Chapter 1, "Why Choose a Career in Infosec?," will answer lots of questions you may already have, including questions about job availability, salary, and whether your personality is built for a career in infosec. If you already know that this is the career path you want, it's still a good idea to read the chapter, because it offers insight into specific pros and cons of this field that you may not have considered.

Even if you're not sure about information security, keep reading, because the next chapter is going to give you some really good information about the industry. It breaks down the many different types of careers within infosec and will be helpful in determining which area you might find most interesting.

Your future awaits! *Getty Images/tortoon*

1

Why Choose a Career in Infosec?

*I*n the introduction, you learned that the infosec/cybersecurity field is growing and is expected to be a strong career path for a long time. At the same
time, you should understand that it's a competitive industry and that to succeed in it, you'll need to keep learning new techniques, languages, and technologies all throughout your career.

This book is not designed to convince you that you should pursue an
infosec career. Instead, its goal is to thoroughly describe various careers within
infosec to help you decide whether it's something you'd like to explore. Infosec
is an interesting career choice, because infosec professionals work in every type
and size of industry and business imaginable—from startups to Fortune 500
companies and beyond.

The good news is that infosec analysts are hot commodities. This means
that they're in great demand, and companies are struggling to fill the slots with
qualified workers. You'll learn more about the demand in the field later in this
chapter.

This chapter discusses five key fields in infosec that were presented in the
introduction. It covers the basic duties and tasks in each. After reading this
chapter, you will have a good understanding of five different types of jobs, and
you can start to determine if one of them is a good fit for you. Let's start with
discussing what infosec professionals do in general.

This book contains a lot of technical jargon. If you run into unfamiliar terms and
descriptions while reading this and other chapters, be sure to check out the detailed
glossary at the end of the book, which defines many of these terms for you.

What Do Infosec Professionals Do?

Network security. Cybersecurity. Internet security. These terms are all related, and they describe different angles of information security (infosec). When you read about large companies and databases being hacked, you know that it was a failure of network security. But for every company whose data is breached in a security incident, thousands of companies keep their data locked up tight, away from prying eyes. This is the job of the infosec analyst.

If you have used a smartphone, purchased something online, entered a password, or even just used a computer, you are probably aware of some of the ways that companies attempt to protect your personal data so that it is secure and inaccessible to others who might use it nefariously. Infosec professionals are those people behind the scenes, working to make networks, databases, clouds, and application systems safe and secure for their users, as well as securing your personal information, tracking down bugs and leaks, and finding and fixing security breaches. In short, infosec workers attempt to secure the many systems that we use every day, as well as our personal data.

Studying and learning new things is a big part of every infosec career. *Getty Images/venuestock*

These specialists analyze and evaluate weaknesses in the infrastructure (this might be the software, hardware, networks, or cloud). They find the best tools and countermeasures to address those vulnerabilities. They may also assess the damage from security incidents and recommend solutions and best practices. They may also test for compliance with security policies and help the organization create, implement, or manage its security solutions.

Infosec analysts learn from each new attack, whether it was carried out on them or on another company. By knowing the tools of the hackers and data thieves, they can help their own company create better defenses. In addition to watching other companies, infosec analysts often attempt to break into their own systems, in an attempt to find holes in their own security systems and patch them before anyone else identifies the vulnerability.

The following is a list of the typical tasks of the infosec analyst:[1]

- Monitor their company's networks/clouds/databases for security breaches
- Be familiar with firewalls and tools like encryption that keep data private
- Report any security breaches or data loss to management
- Attempt to attack their own networks/databases to simulate a hacking attempt
- Recommend security enhancements to management or senior IT staff
- Write security documentation, policies, and procedures for the company's users
- Help the company's employees understand and properly follow the security protocols

So, where do infosec people work? Regardless of the official job title, typical employers of infosec professionals include the following:

- Technology and internet companies
- Security software companies
- Defense companies
- Many government departments and defense/intelligence agencies
- Many IT companies, and IT divisions of companies in many industry sectors
- The e-commerce sector

- Banks, financial firms, and credit card companies
- Really, any medium- to large-sized company!

Infosec has become so integrated into our lives that it's hard to count how many ways it affects us. That's one of the reasons that the industry offers so many different opportunities and different types of jobs. Likewise, the education, certification, and job descriptions for various infosec jobs can differ greatly. This is why it's important to list and describe many different types of jobs in this chapter.

If you enjoy working with people, love to know how things work, are perhaps a little paranoid by nature, and like a fast-paced environment, a career in infosec may be good for you. On the other hand, you may work long hours, and sometimes you'll need to be on call overnight or during the weekend. There can be a lot of stress, because the success of the company rides on the security of its systems. So consider these facets while you learn more.

So, how do you get started in infosec? Even though job demand is very high, it is hard (but not impossible) to enter the field directly, without some experience in IT in general. Entry-level jobs that pave the way for a cybersecurity career can include:

- Systems administrator
- Database administrator
- Web administrator
- Web developer
- Network administrator
- IT technician
- Network engineer
- Computer software engineer
- Programmer

In addition to the five main types of infosec jobs described in this book, see the textbox in the introduction titled "A Rose by Any Other Name." This textbox lists many job titles within infosec that are interchangeable with the titles we use here. For the purpose of consistency, however, we will use the following job titles to discuss the different types of infosec careers thriving in the world today:

- Security analysts/engineers
- Security architects
- Security administrators
- Security software developers
- Cryptographers/cryptologists/cryptanalysts

Another viable and lucrative career in the infosec realm is the chief information security officer (CISO) position. This book doesn't spend as much time describing this job, as it is a high-level management position responsible for the entire information security division/staff. It's not an entry-level position, and it may be something you move to or aspire to after several years working in the field, in the capacity of the other jobs described here.

Security Analysts/Engineers

Security analysts/engineers are the people who analyze and evaluate weaknesses in the infrastructure (could be software, hardware, networks, databases, or clouds used by the company). They find the best tools and countermeasures to address those vulnerabilities. They may also establish best practices. In addition, they may test for compliance with security policies and help the organization create, implement, and manage its security solutions.

These two terms are sometimes even used in different manners. Generally speaking, some say that a security analyst works more on the attack side, performing penetration tests and identifying security issues. A security engineer works more on the defense side, building secure systems and resolving security incidents.[2]

We have listed these here together because there is often overlap between the roles, especially when a security analyst works within an organization (such as on an internal "pentest" team), rather than as an external consultant (often referred to as a security consultant, penetration tester, or ethical hacker).

A pentester (or pen tester) is a professional who performs authorized, simulated attacks on a computer system to evaluate the security of that system and find vulnerabilities in it. They are sometimes also called white hat hackers or ethical hackers.

Security Architects

Security architects design the entirety of the security system or major components of the security system. They may also lead the security design team when building a new security system.

Although they work together closely, a security architect is generally more senior than a security engineer. Security architects set the vision for security systems, and engineers determine how to put it into practice. However, some organizations and recruiters do interchange these terms as well.

More than many other infosec workers, security architects are truly the "big-picture" people within a company or organization. They straddle the fence between the infosec segment of a company and the rest of the organization, making sure they have a good view of security's role within the company, as well as its costs, benefits, and the role it will play in the company's future.

TANYA JANCA: ADVOCATING FOR DIVERSITY AND INCLUSION

Tanya Janca

Tanya Janca is an application security consultant based in Victoria, BC, Canada. She is also known as "SheHacksPurple" and is the founder, security trainer, and coach of https://SheHacksPurple.dev, specializing in training others in software and cloud security. Her obsession with securing software runs deep, from starting her company, to running her own OWASP (Open Web Application Security Project) chapter for years in Ottawa, cofounding a new OWASP chapter in Victoria, and cofounding the OWASP DevSlop open source and education project. With her countless blog articles, workshops, and talks, her focus is clear.

Tanya is also an advocate for diversity and inclusion, cofounding the international women's organization WoSEC, starting the *online #MentoringMonday initiative, and personally mentoring, advocating for, and enabling countless other women in her field. As a professional computer geek of twenty-plus years, she is a person who is truly fascinated by the "science" of computer science.*

Tanya has a book coming out in the fall of 2020, titled Alice and Bob Learn Application Security. *Check it out!*

Can you explain how you became interested in information security as a career path?

I was a software developer for seventeen years when I met an ethical hacker who encouraged me to join the infosec industry. He came into my workplace and gave a "lunch and learn" where he did SQL injection and broke into one of our apps in less than one minute without a username or password. How did he do it? I had to understand this! How do you test for this? How do you prevent this? Through the course of studying this issue, I found other testing opportunities and vulnerabilities. A few weeks later, I went to a buffer overflow workshop. I was hooked!

Then I joined the OWASP (Open Web Application Security Project) community and got even more mentors, locally and internationally too. It's a wonderful community, and those mentors really encouraged me. Everyone was so welcoming. In 2015 it became my formal full-time job, but it was a hobby before then.

In the OWASP community, there were so many people who wanted to help me. I was hired to speak at a conference, and all these people helped me create my first talk; they went through my slides and helped make them better. They gave me endless encouragement. It was at a Meetup and they were so helpful and supportive. Internationally, they are so welcoming and happy to see you. They are glad you are participating and want to help. This profession attracts great people. I have found that other areas of infosec can be less friendly; for example, the pen testers aren't as friendly and can be very competitive. OWASP is much more friendly. I found *my* people at OWASP.

I also started the WoSEC (Women of Security) organization with some women in the industry. I was "crashing" events and inviting all my female friends. We made it into the news. I also invited women to the Python Meetups. We started a club we called "brunch and bitch" once monthly with women in infosec. We had women-only learning circles and workshops. Then a woman in Africa wrote me and asked to start her own chapter. We now have thirty-two chapters around the world and are only about two years old. We "crashed" RSA and DEFCON. There are lots of women leaving STEM, so we are helping to build relationships and connections for jobs, support, etc. I have met so many cool women friends. Support and validation of women is so important in this field. We outreach to universities and colleges too! The message is that women can have a great, multidecade career in infosec.

What's a "typical" day in your job?

My new company is all about application security training. That means I spend a lot of time creating content and doing live streams. I spend time making things and then breaking them. Then I learn from those experiences and create lessons from them.

I also speak for free to give back to the community. For example, I guest lecture about application security at universities and at local Meetups.

I travel a lot to client sites and conferences—where I speak or give workshops or do training in general. While traveling, I visit clients and give formal training there as well. Conferences I speak at include DevOps days, BSides, DEFCON, BlackHat, and DevSecCon.

Topics I might cover include how to weave security through Azure DevOps and what tools to use and why—such as for secure coding. I teach how to launch your app security programs. I help them make sure they are compliant and deal with cloud security. I cover how to monitor the cloud and make sure the experience is safe, secure, and sturdy.

How has the job changed in the last ten years?

People are taking software security seriously now. This is because it costs a lot of money when they don't. When it affects the bottom line, people care. Previously, data breaches were more rare and it didn't matter to the end user—it didn't affect the bottom line so much. End users are a lot more aware now about what it means to have a data breach and they are angry. The industry has started to figure out that a high-quality product must be secure. It's seen as a value now—or conversely, they realize how damaging it can be to your reputation and your bottom line if you experience a breach.

What's the best part of being in this field?

The best thing is that there are so many opportunities. I love learning, and there are always changes and improvements. So many, many jobs and different types of jobs and you can specialize in all sorts of areas—or you can be a generalist. There's great career/job security. There are so many open jobs without a qualified person to do them. You can find jobs anywhere. Once you have some experience, it's quite easy to move around. The first job is hard to find, but after that, it's great. For one, there is no clear career path into infosec. Academia is behind the times in training students to do infosec. They aren't even teaching them to write secure code.

What's the most challenging part of working in this field?

There is not enough clear and concise training for everyone in IT to do their jobs securely. When I was a software developer, I had no idea how to create secure solutions. Stack Overflow was not helpful, for example. All the odds are against software developers because they don't have the support and training they need to do things securely. The industry standard is one hundred developers to each security person, but in my experience, it's way worse than that. You need to teach safety and security when you train software developers. People go to "boot camp" and learn how to code in a week, without any time spent on security. And some of those people get jobs. How is that even legal?! Computer science is so young and application security is also so young, so it's a free-for-all at this point. I am hopeful that people that make programming frameworks will code more safety and security into them. That can help make security more automatic. That is happening at the website level somewhat.

Do you think the current information technology education adequately prepares students to do this job?

Absolutely not. Most postsecondary institutions don't have a specialization for this, and the ones that do, I am not sure what they are covering. Students aren't getting jobs right away. Maybe you will get a brief overview of network security. They don't get into the details. What they learn is usually way behind what the industry is practicing. Part of the reason is that universities don't pay as well as industry, so great teachers don't want to work at the universities. People who work in the industry teach within the industry for way more money *and* they can own their content. The university system is very rigid, and it's not good for the students. And academia doesn't share—you can't own your own content and the university won't share it with any other universities.

Where do you see the field going in the future?

There will be a lot more vendors in this space. It is a $10 billion industry. I see really cool innovations that will make it easier for devs and security teams to get their jobs done—in the way of automation, DevOps (better than Agile or Waterfall methods), and pipelines for building code. Security can get on board too with DevSecOps— we will add our own testing at the same time as they are using DevOps to build their models. Fast security feedback is my dream!

What traits or skills make for a good security analyst?

You must have a love of learning. You learn all the time. You need to talk to and empathize with others. You have to relate to developers and listen to them and not condescend to them. Empathy is really important. Don't be a person who developers want to go around and avoid. If you are not excited to learn new stuff *and* help people, you will hate this job. When it clicks for a developer, that makes me feel so great. If you don't like teaching others (it's a hassle or is annoying to you?), it's not for you.

What advice do you have for young people considering this career?

Learn everything you can and join the local communities. Join OWASP, WoSEC, and DEFCON, etc. There are lots of good people to follow on social media (LinkedIn or Twitter); but be careful, there are jerks too. Read lots! Also, if anyone ever tells you that you can't or you don't belong in infosec or you must take all these certs or put up other blockades, just ignore them and make a note that they give bad advice. There are gatekeepers out there, and you shouldn't follow advice from them. Find the people saying, "You can do it—here's how. Follow me—this way." Sometimes it's jealousy—they did all that stuff and they think you must too. But you can safely avoid and ignore their advice.

How can a young person prepare for this career while in high school?

If you want to work anywhere in infosec, that's really broad. Try to find what you think is coolest, best, and most interesting. If you can, go to a workshop or a Meetup (take a guardian or adult) and find out which area you are interested in. Then find a professional mentor in that area to help you. Try a little bit of all the things. Try to job shadow with people who are doing that job as well. Try to get a co-op term at a tech company as well if you can.

Any closing comments?

If you want to join the infosec field, there will always be a job for you! We really need your help, and we need and want more good people in the field. I hope this book helps you choose a career path in infosec!

Security Administrators

Security administrators typically install and manage organization-wide security systems. They may also do some of the tasks of a security analyst in smaller organizations. They are effectively in charge of installing, administering, and troubleshooting the organization's security solutions. These are the people on the ground, doing the work of security.

Here are some of the specific things that you could be required to do if your title is any of these—security administrator/security architect/security engineer/security analyst:[3]

- Develop security standards and practices
- Create better ways to solve existing production security issues
- Recommend security enhancements to upper management
- Install and use software, such as firewalls and data encryption programs
- Help install or process new security products and procedures
- Scan networks to find vulnerabilities
- Conduct penetration testing
- Monitor networks and systems for security breaches or intrusions
- Install software that notifies users about intrusions
- Develop automation scripts to handle and track incidents
- Test security solutions using industry standard analysis criteria
- Monitor for unusual system behavior
- Supervise changes in software, hardware, and user needs
- Lead incident response activities
- Lead investigations into how breaches happen
- Report findings to management
- Help plan an organization's information security strategy
- Educate staff members on information security through training and awareness
- Recommend modifications in legal, technical, and regulatory areas
- Provide engineering designs for new hardware solutions

In a small company, one person may serve the role of security administrator/security architect/security analyst/security engineer. When systems and companies get larger and more complex, the need arises to break these titles up into more specific and fine-grained titles.

> "This job is so in demand that you can find anything you like in any kind of environment as long as you are willing to move. You are highly transportable and you can always find something that you find interesting!"—Ben Malisow, CISSP, CISM, CCSP, Security+, involved in infosec education for more than twenty years

Security Software Developers

Security software developers develop security software, including tools for monitoring use of software, web/internet traffic analysis, intrusion detection, virus/spyware/malware detection, antivirus software, and so on. They may also integrate and implement security into the applications and software that a company uses to do business.

Experienced security software developers look at software designs from a security perspective in order to identify and resolve security issues. For each phase of the software development lifecycle, they include security analysis, defenses, and countermeasures to build strong and reliable software.

Security software developers could be required to do so or all of the following:

- Create secure software tools and systems with a team of developers
- Provide engineering designs for new software solutions
- Take a lead in software design, implementation, and testing
- Develop a software security strategy
- Implement, test, and operate advanced software security techniques
- Participate in the lifecycle development of software systems
- Design and build prototype solutions
- Have knowledge of attack vectors that may be used to exploit software
- Perform on-going security testing for software vulnerabilities
- Consult team members about secure programming practices
- Research and identify flaws
- Remedy development mistakes
- Troubleshoot and debug issues that arise
- Maintain technical documentation[4]

Security software developers focus more on the software needed to protect a system (such as antivirus software), as well as on continuing to ensure that the software that an organization uses is safe and secure. As you might expect, these two areas could also be separate jobs as well.

Encryption is the practice of encoding information (using a cipher) so that only people who are approved to access it can do so. Unauthorized users cannot access the information without a key that decrypts the information.[5]

Cryptographers/Cryptologists/Cryptanalysts

These professionals use encryption to secure information or to build security software, develop stronger encryption algorithms, and/or analyze the encrypted information to break the code/cipher or to determine the purpose of malicious software. Sometimes they are called digital forensics experts. Because this is a math- and logic-heavy undertaking, cryptographers usually have at least a bachelor's degree in mathematics, computer science, engineering, linguistics, or a similar program.

Cryptology is still widely used by governments and military forces to secure and disseminate secret information, so some jobs might require candidates to hold high-level security clearances. Cryptographers could be required to:[6]

- Analyze large amounts of data from computer programs, viruses, databases, games, and natural occurrences (such as patterns in our DNA) to find a pattern that helps them predict when the next event will occur. If cryptologists can predict when something will occur, they can save money, time, and even lives.
- Decrypt damaging secret cryptocodes and viruses and eliminate them from a system
- Recover data from damaged or erased hard drives
- Gather and maintain evidence
- Work with law enforcement agencies from the police to the CIA to help trace and uncover scams, hackers, and even serial killers
- Create secret codes and cryptograms used to communicate military secrets; protect government, medical, and other private information; and encrypt personal information to protect it from prying eyes on the internet.

Cryptographers create secret codes and cryptograms to secure highly sensitive or personal data.
Getty Images/Jay Yuno

Educational Requirements

For most jobs within infosec, the ideal candidate will have a bachelor's degree
in a field such as computer science, programming, information security, or a
related field. Degrees from other fields, including liberal arts, are acceptable
too, but only if you have a job history that proves your experience within
infosec. Often, infosec analysts have already spent time in other IT roles such
as systems administrator, database administrator, programmer, computer software engineer, or network architect. This additional experience can provide
them with a good perspective about how networks can be used, misused, and
exploited by outside forces.

"To succeed in this field, you should be good at breaking and hacking stuff. Are you
good at picking locks and can spot loopholes in rules and regulations? If you have
that kind of mindset, this might be a good career for you."—Ross Anderson, PhD,
researcher, author, and industry consultant in security engineering

It is possible to get a job in this field with an associate's degree, or even having had only a few classes after high school, but the hurdles to getting that first job are going to be higher. If college is not an option for you, or if you can already show proficiency and knowledge of certain specific support-based concepts and skills without a formal degree, starting off as a technical support specialist or a pen tester might be a good option. This is not a recommendation against a formal education, of course. Bachelor's and master's degrees will always increase your odds of getting a job and increase your potential opportunities. Also, as mentioned, cryptographers usually have at least a bachelor's degree in mathematics, computer science, engineering, linguistics, or a similar program, due to the heavy focus on math and logic.

Some infosec analysts choose to get a master's degree, such as an MBA (master's in business administration). And like most careers in IT, regardless of your initial education level if you choose to be a infosec analyst, it's necessary to continue pursuing additional education throughout your career. Chapter 3 describes the educational paths to consider in much more detail.

Job Outlook and Compensation

There is a multitude of opportunities for anyone interested in the infosec profession. The profession will continue to grow, due to our increasing reliance on networks, clouds, and computers in general, as well as our increasing connectedness with the larger world and the amount of useful and valuable data that is stored electronically. We will continue to need experienced, smart, curious people who want to continue to learn and protect data as threats come and go. In fact, the need for cybersecurity analysts is expected to grow up to 32 percent in the US through 2028. This is more than twice the demand growth for computer occupations in general (13 percent), and four times the expected growth of the average occupation (7 percent).[7]

These statistics show just how promising this career is now and in the foreseeable future:[8]

- *Education:* Bachelor's degree
- *2018 median income:* $98,350 (more than twice the average income for all jobs)

- *Job outlook 2018–2028:* 32 percent (much faster than average)
- *Work environment:* Most infosec professionals work for computer companies, consulting firms, or business and financial companies

WHAT IS A MEDIAN INCOME?

Throughout this book, you'll hear the term "median income" used frequently. What does it mean? Some people believe it's the same thing as "average income," but that's not correct. While the median income and average income might sometimes be similar, they are calculated in different ways.

The true definition of median income is the income at which half of the workers earn more than that income, and the other half of workers earn less. If this is complicated, think of it this way: Suppose there are five infosec specialists in a company, each with varying skills and experience. Here are their salaries:

- $42,500
- $48,250
- $51,600
- $63,120
- $86,325

What is the median income? In this case, the median income is $51,600, because of the five total positions listed, it is in the middle. Two salaries are higher than $51,600, and two are lower. The "average income" is simply the total of all salaries, divided by the number of total jobs. In this case, the average income is $58,359.

Why does this matter? The median income is a more accurate way to measure the various incomes in a set because it's less likely to be influenced by extremely high or low numbers in the total group of salaries. For example, in our example of five incomes, the highest income ($86,325) is much higher than the other incomes, and therefore it makes the average income ($58,359) well higher than most incomes in the group. Therefore, if you base your income expectations on the *average*, you'll likely be disappointed to eventually learn that most incomes are below it.

But if you look at median income, you'll always know that half the people are above it, and half are below it. That way, depending on your level of experience and training, you'll have a better estimate of where you'll end up on the salary spectrum.

The Pros and Cons of Being in Infosec

As you've learned in this chapter, the infosec industry is healthy and poised for growth. Its salaries are generally higher than in other fields, and its expected growth is higher than in many other occupations.

The list of pros of being in infosec is long, but here are some of the highlights. Consider these benefits as you think about whether it's a good fit for you:

- Good salaries are available.
- Massive growth is expected in the industry over the next decade.
- The environment is dynamic and rapidly changing to keep you interested.
- You'll have interesting people to work with.
- There is demand in nearly every type of organization and anywhere you would want to live.
- You'll have the ability to never stop learning.
- You'll have the feeling that you are making a difference.
- The job can be completely remote.
- Traveling opportunities are abundant.
- Because the cybersecurity job cultures vary greatly, you can find one that fits you.
- Many different specialties exist in cybersecurity, which translates to plentiful career and promotion options.

> "One of the best parts of being in this field is when I find people who are part of my tribe and who understand the nuances of what I am talking about. Whether at a conference or teaching a class, anywhere in the world, you'll find people that have the same passion that you have. That's really exciting."—Nadean Tanner, senior manager, Global Technical Education Programs at Puppet (puppet.com)

Not everything about infosec is favorable. Like most industries, it has its share of downsides. The following is a list of cons of working in the infosec profession:

- Long hours
- Hectic pace

- Increasing competition from other countries
- Constant need to stay current with technologies and languages
- Rate of change can make getting older in the profession a challenge
- Lack of support and understanding from some coworkers
- Mentally taxing and challenging
- Lack of diversity within staff

Regarding diversity, members of the industry seem encouraged that the diversity issue is slowly getting better, and that succeeding in IT in general is headed toward more of a merit-focused status. In other words, their hope is that we're headed toward an IT environment in which earning and keeping a good career is based on how well you perform—as opposed to your ethnicity, gender, age, or some other aspect that has nothing to do with job performance.

ADAM SHOSTACK: THREAT MODELING EXPERT

Adam Shostack is a leading expert on threat modeling, as well as a technologist, author, and game designer. He's a member of the BlackHat Review Board, and he helped found the CVE and many other things. He has decades of experience delivering security. His experience ranges across the business world, from founding startups to nearly a decade at Microsoft.

While at Microsoft, he drove the Autorun fix into Windows Update, was the lead designer of the SDL Threat Modeling Tool v3, and created the "Elevation of Privi-

Adam Shostack

lege" game. Adam is the author of Threat Modeling: Designing for Security, *and the coauthor of* The New School of Information Security. *While not consulting, Shostack advises and mentors startups, as a Mach37 Star Mentor and independently, along with holding a number of board and advisory board roles at nonprofits and academic institutions.*

Can you explain how you became interested in information security as a career path?

I was working in a lab doing clinical research with patient data on a network that connected to the internet. This was the early 1990s, and even then we realized that was unsafe and needed to be secured. My boss said yes, you can be responsible and figure out what that security would mean for us.

This was early days of internet connectivity. There was a firewalls mailing list at that time, where famous people talked to regular people about building firewalls. There were no commercial products at the time. I engaged with people on those sites and learned. Usenet news and mailing lists at that time had mostly smart people doing interesting work who were willing to talk to you. For example, there were more people in the room in a small regional conference recently than at my first DEFCON (hacker conference). That really changes the dynamic of the experience and the education.

What's a "typical" day in your job?

I have two kinds of days that are very different.

On one of those days, I work from my home office, talking to customers and prospects via video and conference calls. Might be with clients with problems, writing client-focused deliverables. This is very on my own and introverted.

Other days, I am on-site with clients. Training in a room with ten to five hundred people to try to help them learn the things I've learned and develop their skills. How do they secure the product they are building? Product delivery, done safely. This is very out there and visible. Extroverted.

I prefer to sit at home, but those days I don't accomplish as much. I have developed a set of techniques when I am on the road (exercise, meditation, taking time for nice meals, downtime for myself) to deal best and get through those days, which are less comfortable for me.

How has the job changed in the last twenty years?

How has it *not* changed?! For one, there are now career tracks. When I started, one professor at one university was giving out cybersecurity degrees. You can't go into infosec without certification now. There weren't any vulnerability scanners or secure products. In fact, Dan Farmer got fired from Sun Microsystems for creating SATAN, for example. There was a book of policies that talked about the role of infosec. I was hiring them out of hacker groups. Anyone could get paid to do that work, and it was kind of a novel idea.

What's the best part of being in this field?

Helping to build big systems that are safer and working with product engineers to build really interesting stuff, like self-driving cars, better medical devices, etc. It's emotionally rewarding to help the FDA make medical devices safer.

What's the most challenging part of your job?

My biggest challenge is that we conceal our failures and so we have no idea about what actually works. This has knock-on effects that make our jobs more difficult. A very large majority of break-ins are from phishing. Where should we put our money if we don't know where threats are coming from and why?

Companies hide these issues and what they did to fix them because they don't want to get sued. This is the root of the big problems.

What kinds of challenges does the field impress upon the people who work in it?

There is so much pressure to do and go that people in this field can get burned out. Substance abuse is also a problem in this field. I want to encourage people to put self-care high on the list. Self-care is very important!!

A lot of the problem is the way we teach people as they are coming into the field. They are told that it's always "make or break" and that these problems can be company-killing mistakes. They are told that a security breach can be a company's death knell. But that's wrong. While I was working at Microsoft, it had its fair share of security incidents, yet it's still in business.

This mindset generates so much pressure on people coming into the field. This pressure isn't shared with the larger world. They self-inflect the pressure to fix it all, but they don't have the authority or power to actually do it.

Related to this, we don't have good ways to measure the effectiveness of the things that we do. So we ask our colleagues to use very detailed and drawn-out security methods, but we can't really measure if these methods are working. We don't know if strict passwords even help, for example.

We can't justify the demands that we make. This leads to skepticism from the users, and they don't do what we think is necessary. And then we feel responsible. We create stressors for ourselves and valorize bad behavior, such as excessive drinking. We don't celebrate self-care. We need to recognize those behaviors and choose them not!

Do you think the current education adequately prepares students to enter the job?

This is a hard question for me to answer, because I don't hire a lot of entry-level folks. I hear there are challenges with getting that first job. High school kids should think about learning how to *think* about solving these problems, but also get enough specifics so that they can get a job. It's expensive to get a degree and the certifications. It's important to learn concepts around cybersecurity, which you get with a classical education. The actualities (languages and so on) change fast and are dead quickly. But the *conceptual* education will go forward—how does a compiler work, an OS, a cloud, a network, etc. Certifications and technical skills won't carry you through a whole career. They might carry you to your first job only.

People can learn that big-picture knowledge. Understand why and how things work.

Where do you see the field going in the future?

Fortunately and unfortunately, we live in a time of profound change. Simpler skills (such as security operations center [SOC] analysts) will be replaced by AI systems. The jobs that remain will require more thinking and higher-level skills. You will need these skills to review and manage the AI's tasks, but you will not use those skills yourself. You have to reason about these skills and hold a discussion about them. There will be a challenging discontinuity between the skills learned in school and the skills people are hired for.

In a sense, we will be managing the robots!

What traits or skills make for a good security analyst?

Solid technical grounding, critical thinking, attention to detail, and an ability to build complex mental models of the systems you are working with. Communication skills are also very important.

What advice do you have for young people considering this career?

Well, I started as a systems administrator and then became a consultant at several different startups. When one startup failed, I went to Microsoft for about a decade after that. My point is that I have not had a really planned-out career and that's okay! You can achieve good things without a laser-like plan that you need to follow.

Be sure to balance your focus on the big picture with intense attention to detail and communication. It's easy in this space to wave your hands and say we should do x or y differently, but you need to understand *why* they did that and understand the details of making those changes before you are actually any good. Deep technical understanding must be balanced with higher-level ideas of what to change and why you will or won't be effective.

You need a technical understanding, but you don't have to be a programmer. Tied up in that requirement are issues around lack of diversity and gender roles. I believe you ought to be technical, but you don't have to be a programmer. Being able to program is very valuable because it gives you intuition and empathy, but you don't have to be a programmer. You have to look at code, make small mods, do version control, and so on. Understanding network packets or using a debugger, looking at the state of memory, etc. is the technical stuff I am talking about. If you can't do that, you can't understand the details that make up this world and that will hold you back.

Knowing how to code these days is *almost* the same as addition and subtraction or writing a sentence. When you say, "I am not a math person," you are holding yourself back. These are skills you can learn and are important. You need to know if numbers are reasonable or not when you come across them.

So, don't be afraid of the tech details! Even if you don't use them every day. Programming should be offered as a requirement in schools.

How can a young person prepare for this career while in high school?

Learn to program, learn to read and write critically, practice your five-paragraph essay and your rebuttal essay. Think about how to construct and judge an argument.

Get your hands dirty—build a computer-controlled something on a Raspberry Pi or replace your Roku with a Raspberry Pi. Build things and then learn how to attack them and break them and make them do different stuff. Jailbreak your Kindle and understand how it works. Add programs and learn how it works. Build some things!

Any closing comments?

One: There is a wonderful talk by a guy named Dr. Hamming, called "You and Your Research." It might seem a little old fashioned, but if you want to do important research (or projects/work), you need to know what the important problems are. I rewatch this once a year or so to remind myself. See https://www.youtube.com/watch?v=a1zDuOPkMSw.

Two: When I look back twenty-five years, I see that change is the only constant. That cliché is true. What will jobs in this field look like in twenty years?! I don't know. They will look incredibly different. But critical thinking, communications, not fearing technology—those won't change in twenty years. Spend time honing those skills and you'll be much more prepared for all the change that is guaranteed to happen.

═══════════

Would I Be a Good Security Analyst?

This is a tough question to answer, because really the answer can only come from you. But don't despair: there are plenty of resources both online and else-

Infosec analysts must not only detect data breaches and prevent future ones, but also document where, when, and ideally, by whom the breach took place. *Getty Images/gorodenkoff*

where that can help you find the answer by guiding you through the types of questions and considerations that will bring you to your conclusion.

Of course no job is going to match your personality or fit your every desire, especially when you are just starting out. There are, however, some aspects to a job that may be so unappealing or simply mismatched that you may decide to opt for something else, or equally you may be so drawn to a feature of a job that any downsides are not that important.

Obviously having an ability and a passion for programming, computers, codes and ciphers, as well as investigating and solving riddles and problems in any capacity, are keys to success in this field, but there are other factors to keep in mind. One way to see if you may be cut out for this career is to ask yourself the following questions:

- **Am I a highly curious and tenacious person who enjoys solving problems?**
  You need a real curiosity for the profession and enjoy learning from your mistakes, as well as be open to learning new things. You have to be willing to work at a problem until you find a solution and not easily give up.

- **Am I willing to continue learning at all times and have a passion for infosec?**
  If you can continue to learn from others, you can continue to hone your skill set. The landscape changes rapidly, which means you'll spend a considerable amount of your career learning the latest in infosec. You also have to be passionate about it. It's more of a lifestyle than just a career.
- **Am I a bit paranoid by habit and nature? Do I enjoy breaking things so I can fix them?**
  If you tend not to trust anyone or anything and think a lot about breaking stuff, this might be a perfect profession for you. Many infosec professionals like to think of ways that bad stuff could happen and how they would fix those things.
- **Can I think critically and build complex mental models of the systems I am working with?**
  You need to know what the real attacks are likely to be, and this is possible only if you can see the big-picture view of your systems.
- **Can I consistently deal with people in a professional, friendly way?**
  Communication is a key skill to have in any profession, but particularly here. You need to talk to and emphasize with others. You have to relate to developers and listen to them and not condescend to them. If you don't like teaching others, this is not for you.

Although it's one thing to *read* about the pros and cons of a particular career, the best way to really get a feel for what a typical day is like on the job and what the challenges and rewards are is to talk to someone who is working in the profession. You can also learn a lot by reading the interviews with actual cyberanalysts that you find sprinkled throughout this book.

If the answer to any of these questions is an adamant no, you might want to consider a different path. Remember that learning what you *don't* like can be just as important as figuring out what you do like to do.

Summary

Most of this chapter was devoted to describing the different types of infosec jobs and a little bit about them:

- Security analysts/engineers
- Security architects
- Security administrators
- Security software developers
- Cryptographers/cryptologists/cryptanalysts

The chapter went into detail on each job type, including example jobs within each category, the job outlook, whether demand for that job is expected to grow in the near future, the educational requirements, and the compensation you can expect.

One important point to remember from this chapter is that even though infosec might technically be considered a single field or industry, there are many diverse types of jobs that are considered "infosec jobs" in nearly every type of business imaginable. So if there is any industry in which your opportunities are nearly limitless, infosec is probably it.

Chapter 2 explores how to build a plan for your future. It discusses everything from educational requirements and certifications to internship opportunities within the infosec industry. You'll learn about finding summer jobs and making the most of volunteer work as well. While infosec jobs are numerous, the industry is quite competitive. This chapter will discuss how you can set yourself apart from the crowd.

2

Forming a Career Plan

*N*ow that you have some idea what infosec is all about, and maybe you even know which branch of it you are interested in, it's time to formulate a career plan. For you organized folks out there, this can be a helpful and energizing process. If you're not a naturally organized person, or perhaps the idea of looking ahead and building a plan to adulthood scares you, you are not alone. That's what this chapter is for.

After we talk about ways to develop a career plan (there is more than one way to do this!), the chapter dives into the various educational requirements. Finally, we will look at how you can gain experience in your community. Yes, experience will look good on your resume and in some cases it's even required. But even more important, getting out there and working within various settings is the best way to determine if infosec is really something that you enjoy. When you find a career that you truly enjoy and have a passion for, it will rarely feel like work at all.

If you still aren't sure if this field is right for you, try a self-assessment questionnaire or a career aptitude test. There are many good ones on the web. As an example, the career-resource website monster.com includes its favorite free self-assessment tools at https://www.monster.com/career-advice/article/best-free-career-assessment-tools. The Princeton Review also has a very good aptitude test geared toward high schoolers at www.princetonreview.com/quiz/career-quiz.

This chapter could just as well have been titled "How to Not End Up Miserable at Work." Because really, what all this is about is achieving happiness. After all, unless you're independently wealthy, you're going to have to work. That's just a given. If you work for eight hours a day, starting at eighteen and retiring at sixty-five, you're going to spend around one hundred thousand hours at work. That's about eleven years. Your life will be much, *much* better if you find a way to spend that time doing something you enjoy, that your personality is suited for, and that your skills help you become good at. Plenty of people don't get to do that, and you can often see it in their faces as you go about your day interacting with other people who are working. In all likelihood, they did not plan their careers very well and just fell into a random series of jobs that were available.

> "If you want to work anywhere in infosec, that's really broad. Try to find what you think is coolest, best, and most interesting. If you can, go to a workshop or a Meetup (take a guardian or adult) and find out which area you are interested in. Then find a professional mentor in that area to help you. Try a little bit of all the things. Try to job shadow with people as well. There are lots of people in the industry who will be glad to help you."—Tanya Janca, founder, security trainer, and coach of SheHacksPurple.dev, specializing in training others in software and cloud security

So, your ultimate goal should be to match your personal interests/goals with your preparation plan for college/careers. Practice articulating your plans and goals to others. Once you feel comfortable doing this, that means you have a good grasp of your goals and the plan to reach them.

What Are You Like?

Every good career plan begins with you. A good place to start is by thinking about your own qualities. What are you like? Where do you feel comfortable and where do you feel uncomfortable? Ask yourself the questions in the textbox titled "All about You" and then think about how your answers match up with the infosec career.

ALL ABOUT YOU

Personality Traits
- Are you introverted or extroverted?
- How do you react to stress—do you stay calm when others panic?

- Do you prefer people or technology and machinery?
- Are you more creative or more analytical?
- Are you better at making things or explaining things?
- Are you organized or creative, or a little of both (or neither)?
- How much money do you want to make—just enough or all of it?
- What does the word "success" mean to you?

Interests

- Are you interested in how things work?
- Are you interested in solving problems?
- Are you interested in helping people?
- Are you interested in moving up a clear career ladder?
- Or would you like to move around from one kind of job to another?

Likes and Dislikes

- Do you like to figure things out or to know ahead of time exactly what's coming up?
- Do you like working on your own or as part of a team?
- Do you like talking to people, or do you prefer minimal interaction?
- Do you like to figure out problems and solve them?
- Can you take direction from a boss or teacher, or do you want to decide for yourself how to do things?
- Do you like things to be the same or to change a lot?

Strengths and Challenges

- What is something you accomplished that you're proud of?
- Are you naturally good at school, or do you have to work harder at some subjects?
- Are you physically strong and active or not so much?
- Are you flexible and able to adapt to changes and new situations?
- Are you better at math or better at English?
- Are you better at computers or doing things with your hands?
- What is your best trait (in your opinion)?
- What is your worst trait (in your opinion)?

Remember, this list is only for you. You're not trying to impress anybody or tell anyone what you think they want to hear. You're just talking to you. Be as honest as you can—tell yourself the truth, not what you think someone else would want the answer to be. Once you've got a good list about your own interests, likes and dislikes, strengths, and challenges, you'll be in a better position to know what kind of career you want.

ABOUT THE JOB

What kind of work will you be doing?

- What kind of environment will you be working in?
- Will you have regular 9-to-5 hours, or will you be working evenings, weekends, and overtime?
- What kind of community would you be living in—city, suburb, or small town?
- Will you be able to live where you want to? Or will you need to go where the job is?
- Will you work directly with customers?
- What will your coworkers be like?
- How much education will you need?
- Do you need certification?
- Is there room for advancement?
- What does the job pay?
- What kind of benefits will the job provide (if any)?

Is there room to change jobs and try different things?

Planning the Plan

In this chapter, you are on a fact-finding mission of sorts. A career fact-finding plan, no matter what the field, should include these main steps:

- Take some time to consider and jot down your interests and personality traits, with the previous sections' help.

- Find out as much as you can about the day-to-day of infosec specialists at all levels. In what kinds of environments do they work? Who will you work with? How demanding is the job? What are the challenges? Chapter 1 of this book is designed to help you in this regard.
- Find out about educational requirements and schooling expectations. Will you be able to meet any rigorous requirements?
- Seek out opportunities to volunteer or shadow others doing the job. Use your critical thinking skills to ask questions and consider whether this is the right environment for you. This chapter also discusses ways to find job-shadowing opportunities and other job-related experiences.
- Look into student aid, grants, scholarships, and other ways you can get help to pay for schooling. It's not just about student aid and scholarships, either. Some larger organizations will pay employees to go back to school to get further degrees.
- Build a timetable for taking requirements exams such as the SAT and ACT, applying to schools, visiting schools, and making your decision. You should write down all important deadlines and have them at the ready when you need them.
- Continue to look for employment that matters during your college years—internships and work experiences that help you build hands-on experience and knowledge about your actual career.
- Find a mentor who is currently working in your field of interest. This person can be a great source of information, education, and connections. Don't expect a job (at least not at first); just build a relationship with someone who wants to pass along their wisdom and experience. Coffee meetings or even emails are a great way to start.

The whole point of career planning is not to overwhelm you with a seemingly huge endeavor; it's to maximize happiness. Your ultimate goal should be to match your personal interests/goals/abilities with your preparation plan for college/careers. Practice articulating your plans and goals to others. Once you feel comfortable doing this, that means you have a good grasp of your goals and the plan to reach them.

A mentor can help you in many ways. *Getty Images/filadendron*

YOUR PASSIONS, ABILITIES, AND INTERESTS: IN JOB FORM

Think about how you've done at school and how things have worked out at any temporary or part-time jobs you've had so far. What are you really good at, in your opinion? And what have other people told you you're good at? What are you not very good at right now, but you would like to become better at? What are you not very good at, and you're okay with not getting better at?

Now forget about work for a minute. In fact, forget about needing to ever have a job again. You won the lottery—congratulations. Now answer these questions: What are your favorite three ways of spending your time? For each one of those things, can you describe why you think you in particular are attracted to it? If you could get up tomorrow and do anything you wanted all day long, what would it be? These questions can be fun, but they can also lead you to your true passions. The next step is to find the job that sparks your passions.

Where to Go for Help

If you aren't sure where to start, your local library, school library, and guidance counselor office are great places to begin.

VISIT THE GUIDANCE COUNSELOR OFFICE

Each school and each state handles things a little differently, but a high school guidance office will usually have several resources for you, including help for applying to college. Some of the resources you might find in your high school guidance office include the following:

- Seniors handbook
- Higher education handbook
- College visit forms
- Student brag sheet/resume forms
- Useful links for college planning and searches
- Useful links for scholarships/financial resources
- NCAA hints for sports scholarships
- SAT and ACT information
- Common App tips and application
- Essay-writing tips

Your guidance office can also help you take interest and aptitude tests, so you can narrow down your interests or discover new ideas you've never thought of. They can help you develop a resume and a career portfolio to show what you've done and what you're capable of. Your guidance counselor can work with you individually; they also put on school-wide or grade-wide workshops, classes, focus groups, or presentations about job skills and personal development.

Make an appointment with a counselor or email them and ask about taking career interest questionnaires. With a little prodding, you'll be directed to lots of good information online and elsewhere.

DO THE RESEARCH YOURSELF

Of course, a well-stocked guidance office or school library/media center will have books like this one for you to explore! You can also start your research with these four sites:

- The Bureau of Labor Statistics' Career Outlook site at www.bls.gov/careeroutlook/home.htm. The US Department of Labor's Bureau of Labor Statistics site doesn't just track job statistics, as you learned in chapter 1. There is an entire portion of this site dedicated to young adults looking to uncover their interests and match those interests with jobs currently in the market. There is a section called "Career Planning for High Schoolers" that you should check out. Information is updated based on career trends and jobs in demand, so you'll get practical information as well.

- The Mapping Your Future site at www.mappingyourfuture.org helps you determine a career path and then helps you map out a plan to reach those goals. It includes tips on preparing for college, paying for college, job hunting, resume writing, and more.

- The Education Planner site at www.educationplanner.org has separate sections for students, parents, and counselors. It breaks down the task of planning your career goals into simple, easy-to-understand steps. You can find personality assessments, get tips for preparing for school, learn from some Q&As from counselors, download and use a planner worksheet, read about how to finance your education, and more.

- TeenLife at www.teenlife.com calls itself "the leading source for college preparation," and it includes lots of information about summer programs, gap year programs, community service, and more. They believe that spending time out "in the world" outside of the classroom can help students do better in school, find a better fit in terms of career, and even interview better with colleges. This site contains lots of links to volunteer and summer programs.

Use these sites as jumping-off points, but don't be afraid to reach out to a real person, such as a guidance counselor or your favorite teacher, if you're feeling overwhelmed.

TALK TO PROFESSIONALS IN THE FIELD

After you've done some research on your own, you'll already have an idea of which infosec field appeals to you most. But what is it like to really do that job?

One of the best ways to learn what a job is like is to talk to someone who does it. Start with your own city or town. Infosec professionals are found in every business, university, governmental body, and nonprofit organization where you live. Ask people you know for introductions to people they know. Or just look them up online and contact them yourself. Even if those people are farther away, there are ways to see these careers through their eyes. Consider these avenues to connect with professionals in the field:

- *Informational interviews:* These can be in person, on the phone, or online via Skype, Zoom, or a similar program. Ask to speak to the person for twenty or thirty minutes. It's important to respect that they are likely to be very busy and it may be difficult for them to spare you even that much time. Ask them open-ended questions about the job itself, how they chose it, what they like and don't like. Be sure to ask the two most important questions at the end:
  - What other advice would you give me?
  - Who else should I talk to?
- *Job shadowing:* This is an opportunity to spend a day with a professional in order to learn about a career and observe daily work activities. This kind of program is usually organized by your high school guidance office or sometimes through a community program like Junior Achievement.
- *Summer internship programs:* These opportunities can provide a high school student with valuable professional development, mentoring, and job shadowing alongside hands-on work. Check out the "Resources" section in this book for more information about internships for high school students interested in infosec professions.

BEN MALISOW: TRAINER, WRITER, AND SPEAKER

Ben Malisow, CISSP, CISM, CCSP, Security+, has been involved in infosec education for more than twenty years. At Carnegie Mellon University, he crafted and delivered the CISSP prep course for CMU's CERT/SEU. Malisow was the ISSM for the FBI's most highly classified counterterror intelligence-sharing network, served as a US Air Force officer, and taught grades six through twelve at a reform school in the Las Vegas public school district (probably his most dangerous employment to date). He has also written several books on security, including the CCSP ISC² Certified Cloud Security Professional Official Study Guide *and* How to Pass Your INFOSEC Certification Test: A Guide to Passing the CISSP, CISA, CISM, Net- work+, Security+, and CCSP. *In addition to other consulting and teaching, Ben is a certified instructor for ISC², delivering CISSP and CCSP courses. You can reach him at www.benmalisow.com.*

Ben Malisow

Can you explain how you became interested in information security as a career path?

I was always interested in securing people and things. I got into the military right after high school. First it was physical security; then I moved to information security while in the military. I was a command and control officer and responsible for oper- ational security, which is trying not to reveal sensitive information by giving out too much nonsensitive info. This reduces opportunity for aggregation. Training people not to talk about stuff they shouldn't be sharing.

From very early on, I loved training as well. Many folks can do the job but aren't good at talking about it. But I can talk about it. I like talking in front of groups. Soon out of the military, I got into the federal contracting space and started working for DARPA (Defense Advanced Research Project Agency). I built and delivered their training program for several years. Then I went into community college teaching, was a high school teacher, and taught at the university level. Now I teach for ISC² and do private classes and write books.

What's a "typical" day in your job?

I do a lot of training at this point, which I enjoy.

A new person coming into the field will likely come in and start your day by reviewing everything from the previous shift/day. You see what's developed and if there are new alerts. Then you look forward. You browse databases and informational streams for what's happening in the industry—such as LinkedIn or even reading classified intelligence feeds to determine what new threats are being projected by analysts. You try to stay ahead. Lots of meetings where you report to management and the business unit. You work on projects such as updating and upgrading systems and processes. Configuration management—tracking and tabulating everything. Incident response also. You might just support or report to incident response. This all depends on your position and area of infosec you work on.

There is so much specialization now! There are many niches and you may see only a small piece. Niches include incident response, network monitoring teams (as part of a security operation center), monitoring network performance as well as intrusion or data loss, as well as forensics (although not many organizations have permanent teams). Usually forensics is a specialized third-party provider.

There is also the policy and governance side, which is the decision-making process to determine how much risk the organization can handle, costs of this, and how to handle and avoid these risks—this is about writing policy. There is also audit and compliance—they review and see that you are following rules properly. This involves lots of attention to detail—accounting people overlap with this.

There is also network intrusion monitoring to see if someone tried to break in. System administration can even have a security component—maintaining the current build with optimum security and ensuring secure software development, where app developers work to make sure the app is secure before it goes into production. Security architects should understand how all the pieces go together so the overall environment is secure. There is a lot of overlap today between physical and info security.

How has the job changed in the last twenty years?

For one, there are now dedicated schools, classes, and training programs to get into security. People like me came in through other means—accounting, general IT, economics, or physical security. Computer science classes were still fairly new and evolving. No one had a security degree. Now, young people can go straight into college and study it in college.

I think it's great that security is more prominent. Productivity used to be the main thing. After 9/11, physical security became a major concern, especially in the US. In recent years, privacy has also become a major concern, especially in Europe.

I don't think all these rules and regulations are necessarily making the world better or more optimized. Individuals are not more secure. The global environment requires organizations to follow a lot of rules that may not apply to them and don't really help their processes, simply because someone passed a law somewhere. But security is not a one-size-fits-all solution. It's not the best we could be doing. It's part of a growth thing and we are chasing the past. Government moves slowly and laws are always behind the pace of technology. Also, the law can simply be misapplied.

I also think, from the individual/user standpoint, the things they *think* they want are not always the things they really want. Privacy is a good example. Users say it's important but then post personal photos, oblivious to what the info means and how much of it there is. Users don't realize how this behavior could negatively affect them.

What's the best part of being in this field?

I get to meet many different people. I travel a lot and the job is very varied. I love the different places and different people. Not true for all infosec people, but it is for me. The job is so in demand that you can find anything you like in any kind of environment as long as you are willing to move. You are highly transportable, and you should find something that you find interesting. Think of all the different, cool industries that need security! Motion picture industry, gaming, and so on. All industries use security personnel.

What's the most challenging part of your job?

Staying current for sure! I am a dinosaur and simply being aware of stuff is hard enough, but understanding them all is next to impossible. As those niches grow, just being familiar with them all is very difficult. Young people raised in these niches will be in the know.

You can't know something for any length of time and achieve mastery. It will move past you. Programming languages, etc. all become obsolete.

Do you think the current education adequately prepares students to do this job?

No. We give a lot of good foundational background. It mostly signals that you can get a four-year degree. I want people with a degree who also have worked in the field and have experience. The academic work and the practical work are very different.

Where do you see the field going in the future?

One of two ways—it will continue to be very specialized and rewarding in the future. The laws are continuing to go that way. Or, it could go away, which could lead to a less private existence. Informational parity, which would lead to no security. But I don't see that happening anytime soon.

What traits make for a good security analyst?

Paranoid by habit, nature, and trade. You don't trust anyone or anything. That's an unusual creature! You think a lot about breaking stuff. Kind of weird. You like to think of ways that bad stuff could happen (but you don't actually do the bad stuff). You also need to have lots of attention to detail, lots of patience, and be able to perform repetitive tasks with attention.

What advice do you have for young people considering this career?

Degrees and certifications are great, but I highly recommend people getting actual IT experience doing anything. Get your hands dirty. Get real experience doing anything you can. Demonstrate that you can use the knowledge you learn and start as early as possible. Work at a help desk, for example. Don't specialize in any one technology or vendor and don't know only a single coding language, but do understand the overall systems and disciplines. Are you adaptable, dependable, and trustworthy? It may mean taking low-paying jobs at first, but the learning potential is great. You can do that in concert with going to college.

Also, I am biased, but I like professional certifications and vendor-agnostic certifications over degrees. For example, a general cloud certification is better, in my opinion, than one specifically from Cisco. With the latter, you are trained on one tool. But can you promise that it is always the best tool for the job? You need a big-picture approach to determine what the best tool for the job is.

How can a young person prepare for this career while in high school?

Be a nerd! Read everything you can. Practice on your system. Don't be afraid to break things that you own. Experiment and try. Find the free videos. Good info is out there—go and grab it!

Any closing comments?

Whatever it is you are good at, try to use that as your strength and see how it fits into the industry if that's what you want to do. For instance, I am good at writing and talking about the industry, and that found me a place in infosec. You can find a way to make it work with your unique talents. There is such a wide breadth of options, and you can use your talents to find a way in.

Making High School Count

Regardless of the career you choose, there are some basic yet important things you can do while in high school to position yourself in the most advantageous way. Remember—it's not just about having the best application; it's also about figuring out which areas of infosec you actually would enjoy doing and which ones don't suit you. Consider these steps toward becoming a well-rounded and marketable person during your high school years:

- Volunteer at your high school or local library's help desk.
- Take as many programming and computer design and skills classes as possible during your high school career.
- Take intro classes in logic and discrete mathematics.
- Take online computer courses to enhance and expand your breadth of knowledge. You can find many free courses online.
- Use the summers to get as much experience working with computers as you can. Be comfortable using all kinds of computer software.
- Learn first aid and CPR. You'll need these important skills regardless of your profession.
- Hone your communication skills in English, speech, and debate. You'll need them to speak with everyone, from coworkers to clients and bosses.
- Consider getting certified in your areas of interest, if you have the time and money to do so.
- Volunteer in as many settings as you can.

CLASSES TO TAKE IN HIGH SCHOOL

High school is a good time to take as many electives and special-interest classes as you can because doing so will give you a feeling for what you like and don't like, and it will give you experience that you can use as a stepping-stone to find internships and other positions. If your high school is on the small side, you might not have access to all the options listed here, but take what you can. It will build your portfolio and help you discover where your passions lie within the infosec field.

- Basic word processing
- Computer programming

- Computer animation
- Application development
- Logic and discrete mathematics
- Computer repair
- Graphic design
- Media technology
- Video game development
- Web design
- Web programming
- Cybersecurity

Taking many classes offered by the computer science/information technology department will also give you the opportunity to get to know the instructors who teach those classes. These are people who have likely worked, or currently work, in the fields they teach, which means they have connections and can teach you about the field outside of the classroom as well. Note that many colleges look for comments or recommendations from teachers when considering an application. Building good relationships with your teachers can be a great way to improve your chances of positive recommendations.

Educational Requirements

Most people in the IT field (programmers, web developers, systems analysts and support, network engineers, security analysts and support, etc.) enter the field with a bachelor's degree (four years) in something resembling computer science (information science, information technology, programming, etc.). However, it is also possible to be successful in the field with an associate's degree (two years), especially if you are well versed in multiple programming languages and/or are certified. The following sections cover the traditional educational requirements in detail, which includes common certifications that people working in cybersecurity are expected to have.

Typical Degree Paths

If you want to have a successful career in the infosec field, it's best to have at least a bachelor's degree in a relevant field[1] (discussed soon). You'll be smart to pursue a bachelor's degree in a computer-related field (such as information technology or computer science) or in engineering (electrical engineering is preferable). More and more universities are now offering a four-year degree specifically in cybersecurity as well. Be sure to check out chapter 3, which covers the search for the right higher-ed education in more detail.

Bachelor's programs usually contain some beginner classes that will help you learn about the cybersecurity basics. These essentials will help you be prepared for a job after graduation and will help you decide what area of cybersecurity you like best.[2]

Having a master's degree will most likely place you into a better entry-level position, which will likely include a higher starting salary. However, with a few years of experience, the difference between a bachelor's and master's degree won't matter as much as whether and how well you've kept up with changes in your field.

Keeping your knowledge, education, and certifications up to date is extremely important. *Getty Images/ monkeybusinessimages*

You must have the motivation and desire to continue to learn and be educated and certified as things evolve. These are fast-paced, always changing fields, and you will need to stay on top of the latest trends and changes in order to be marketable and stay relevant, whether that be in terms of learning new programming languages, understanding how to avoid and mitigate the latest hacks and security threats on your systems, or simply keeping up with updates and version changes in the system that you support. We touch on some of these typical certifications shortly, but since these change often, you should research your area of interest to find out what's happening *right now* in terms of certifications.

The part of your education that is extremely important and specific to your career path involves the certifications, program/system/network knowledge, and know-how you have.

Certification Options in Systems and Security Analysis

As mentioned, being certified and keeping those certifications up to date during the course of your career is extremely important when you work in cybersecurity. Table 2.1 lists the main infosec certifications at the time of this writing that employers look for when evaluating prospective candidates. These certifications are considered to be leaders in the field of information security at this time.

At this point in your career search, it could be helpful to investigate these certifications to see if you find the topics they cover interesting, feel that you could learn and understand them with schooling, and could imagine working with these topics as a career.

Most, if not all, of these certifications require an applicant to be a minimum of eighteen years of age. If you are a few years off, there is still lots you can do to learn and prepare! Take basic networking courses; then create a home network and add as many elements as you can to it. Learn everything you can about systems and networks before you start messing with security. Be curious about everything and see what happens when you use parameters for configurations that aren't "in the books." Get a good grounding in basic statistics and learn at least one scripting language (such as Python).

Table 2.1. Main Infosec Certifications

Certification Name	Description	Coverage	Granter/Certification Body
Certified Ethical Hacker (CEH)	Teaches the skills needed to think and act like a hacker.	Hacking technologies that target cloud computing technology, mobile platforms, and the latest operating systems, latest vulnerabilities, malware and viruses, and infosec laws and standards.	International Council of Electronic Commerce Consultants (EC-Council)[1]
CompTIA Security+	A base-level certification for IT professionals new to cybersecurity. Regarded as a general cybersecurity certification because it doesn't focus on a single vendor product line.	Broad IT security concepts, such as network attack strategies and defenses, threat analysis, business continuity and disaster recovery, and encryption standards and products.	CompTIA[2]
Certified Information System Security Professional (CISSP)	This certification is not vendor specific, so the knowledge can be applied to various setups.	Security-related issues in access control, telecommunications, and networking. Many IT companies consider CISSP a base requirement for employees responsible for network security.	International Information System Security Certification Consortium (ISC)[2][3]
Certified Information Security Manager (CISM)	Advanced certification for analysts with several years of experience.	Covers infosec program development and management, infosec management, infosec incident management, and information risk management and compliance.	ISACA, a nonprofit credentialing organization[4]
Certified Information Systems Auditor (CISA)	The main requirement for high-level infosec audit, assurance, and control positions.	Focuses on information auditing. Topics include the process of auditing information systems, IT management and governance, and protecting information assets.	ISACA, a nonprofit credentialing organization

Certification	Description	Issuing Organization
NIST Cybersecurity Framework (NCSF)	Teaches the NIST Cybersecurity Framework, which was created to provide a uniform standard that government and businesses could adopt to guide their cybersecurity activities and risk management programs.	NIST (National Institute of Standards and Technology), US Department of Commerce[5]
	Validates that cybersecurity professionals have the baseline skills to design, build, test, and manage a cybersecurity program using the NIST Cybersecurity Framework.	
Certified Cloud Security Professional (CCSP)	All about cloud computing and its information security risks and mitigation strategies.	International Information System Security Certification Consortium (ISC)[2]
	Covers cloud architecture and design, cloud data security, cloud platform and infrastructure security, cloud security operations, and more.	
Computer Hacking Forensic Investigator (CHFI)	Teaches identifying an intruder's footprints and properly gathering the necessary evidence to prosecute in a court of law.	International Council of Electronic Commerce Consultants (EC-Council)
	Geared for e-business security professionals, systems administrators, IT managers, law enforcement, and defense and military personnel.	

Notes

"The Best Cybersecurity Certifications to Boost Your Career in 2018," New Horizons Computer Learning Centers, https://www.newhorizons.com/article/the-best-cybersecurity-certifications-to-boost-your-career-in-2018.

1 "Certified Ethical Hacker Program," EC-Council, https://www.eccouncil.org/programs/certified-ethical-hacker-ceh/.

2 "CompTIA CySA+," CompTIA, https://www.comptia.org/certifications/cybersecurity-analyst.

3 "CISSP—The World's Premier Cybersecurity Certification," (ISC)², https://www.isc2.org/Certifications/CISSP.

4 "Certifications," ISACA, https://www.isaca.org/credentialing/certifications.

5 "Cybersecurity Framework," National Institute of Standards and Technology, US Department of Commerce, https://www.nist.gov/cyberframework.

"It's important to learn concepts around cybersecurity, which you get with a classical education. The actualities (languages and so on) change fast and are dead quickly. But the *conceptual* education will go forward—how does a compiler work, an OS, a cloud, a network, etc. Certifications and technical skills won't carry you through a whole career. They might carry you to your first job only. Spend time honing your critical thinking and communication skills first."—Adam Shostack, noted expert on threat modeling, as well as a technologist, author, and game designer

Large or Small Company?

What kind of company do you want to work for? And how do you decide? There are opportunities for infosec professionals at every scale.

Large companies and small companies have both similarities and differences. Whether those qualities are pros or cons depends on you and what you want out of your work experience. Let's compare a few differences.

LARGE COMPANIES

Large companies have the following pros and cons:

- *Structure:* Larger companies tend to have more organization in their organization. There's probably a clear hierarchical structure as well as a clear career ladder.
- *Your role:* In a larger company, your job is likely to be well defined. You'll know what is expected of you and where you should be focused.
- *Benefits:* Larger companies sometimes have better benefits (health insurance, dental plans, etc.) because having a large pool of employees puts them in a better bargaining position with insurance companies. You are more likely to find a range of insurance options to choose from.
- *Salaries and bonuses:* Larger companies may be able to offer higher salaries and annual or target-based bonuses because they have more money than smaller companies.
- *Location:* When you work for a larger company, be prepared to be sent where they need you. You could end up anywhere in the country or in the world!

- *Training and educational benefits:* Larger companies often have extensive training programs or pay for most or all of your continuing education credits.
- *Other opportunities:* A larger company may have more types of jobs, meaning that you can move "sideways" into a different type of work, as well as moving up the ladder in your original career track.

SMALL COMPANIES

Small companies have the following pros and cons:

- *Structure:* Smaller companies are often more flexible than larger ones. This means they can respond to changes in the market quickly. When a nimble company pivots, you can pivot with it.
- *Your role:* In a smaller company, your job may grow and change more easily than in a larger company.
- *Benefits:* Smaller companies may offer only one (or no) health insurance option, and may offer fewer benefits than a larger company. But they are in a better position to be flexible when you have something unexpected come up—and that will depend more on your boss's attitude than on the rules a larger company would have to apply.
- *Salaries and bonuses:* Smaller companies may not be able to offer the highest salaries or bonuses.
- *Location:* When you work for a smaller company, you are most likely to be working where the company is based. That means you have a better chance of staying in your own community, if that's an important factor for you.
- *Training and educational benefits:* Smaller companies don't usually run their own training programs, but they do offer apprenticeships with seasoned workers. Some small companies will pay for training and continuing education at nearby community or technical colleges and trade schools.
- *Other opportunities:* With a smaller company, you have a better opportunity to see how a business is run, to learn about the organizational and management side of the company, which is very useful if you might want to be a boss yourself one day.

Making the Most of Your Experiences

As mentioned earlier, experiences you gain out in the real world are very important pieces of the puzzle that include your education. In other words, don't just use the time you spend in the classroom learning C++ or Python, for example, to determine if you have picked the right career path. Getting out in the real world, perhaps working in an office, will help you determine what you do and don't like and what kind of shape you want your career to take. It doesn't hurt your resume and college applications either!

It can be hard to get your foot in the door, though. How do you get experience in the real world without having a degree or any experience to begin with? Try these approaches to gaining some much-needed experience:

- Volunteer at your high school or local library's help desk.
- Volunteer in the IT department of a charitable or nonprofit organization that you support and ask them for a professional reference moving forward.
- Write a blog in the technical area where you want to get experience. Become an active contributor on technology-related discussion boards.
- Get a job as a troubleshooter, such as at Best Buy in the Geek Squad.

Be sure to ask for lots of feedback from your mentors, bosses, and from others. Don't be afraid to ask for advice, and then be sure to have an open attitude about the information you get back.

Building Your People Skills

The ability to get along with people is one of the three most important factors of a successful career in any field. You'll be interacting with lots of people—clients, coworkers, supervisors, and others. The ability to communicate well—both orally and in writing—is essential to getting a job, keeping a job, and doing that job well.

There are some basic people skills that anybody can learn—and that are useful for everyone, regardless of their career plans. Let's take a look at some of those important skills.

RELATING TO OTHER PEOPLE

This can be summed up as treating others the way you would want them to treat you. For instance:

- Try to see the other person's point of view (even if it's different from yours). Empathy and compassion go a long way.
- Be understanding and respectful toward other people.
- Be patient—nobody is perfect (including you).
- Pay attention and show genuine interest—everyone has something interesting about them. Take the time to find out what it is!

COMMUNICATION SKILLS

Good communication is essential in work and in life. Basic communication skills include:

- *Active listening:* Paying attention to what the other person is actually saying and responding to it (not just planning what you're going to say next)
- *Speaking:* Expressing yourself simply and clearly in spoken words
- *Writing:* Expressing yourself simply, clearly, and with sufficient detail in writing. Avoid a lot of extra words that can confuse the reader, but don't leave out anything important.
- *Body language:* Nonverbal communication is just as important as verbal communication. Pay attention to the messages you're sending with your face, gestures, and posture, and pay attention to the nonverbal messages you're receiving from the people around you.

CHARACTER

Your character includes your personality and the choices you make based on your values and beliefs. Important character traits include:

- *Honesty and trust:* Honesty and trust form the basis of any relationship, whether personal or professional. This includes trusting others as well as being trustworthy yourself. When you and your coworkers know you can count on each other, you can accomplish anything. And when clients know they can trust you and the work you do for them, you will be set up for a lifetime of success.

- *A sense of humor:* Knowing when to use humor to lighten a situation is a great people skill. Used appropriately, humor can defuse tension and conflict. And be able and willing to laugh at yourself.
- *Being supportive and helpful:* Offer to do a little more than required or to help someone out when they need it. If someone is having a hard day, be respectful of their feelings.
- *Flexibility:* Be ready to adapt to changing situations, conditions, and workflow.
- *Good judgment:* Choose your own behavior—don't just go along with something if your gut says it's not a good idea.

When you get to chapter 4, you'll learn more about putting your people skills into action to get and keep the job you want.

Networking

Because it's so important, a last word about networking. It's important to develop mentor relationships even at this stage. Remember that as much as 85 percent of jobs are found through personal contacts.[3] If you know someone in the field, don't hesitate to reach out. Be patient and polite, but ask for help, perspective, and guidance.

There are basically two kinds of networking:

- *Internal networking:* This involves reaching out to people you already know, such as at your internship or at school. These people don't necessarily have to work in infosec. They may have other advice or ideas that will help you on your journey. Be sure to give back, too. You don't want to be the one who is always asking for help but never giving any! Take care of these relationships. They are valuable in too many ways to list.
- *External networking:* This involves meeting new people at work, in extracurricular clubs, in student chapters of professional associations, at conferences or workshops, or anywhere that you don't already spend a lot of time. If you discover someone you'd like to know or ask a ques-

tion, seek them out and introduce yourself. Be polite and professional. Don't take up too much of their time, at least at first.

If you don't know anyone, ask your school guidance counselor to help you make connections. Or pick up the phone yourself. Reaching out with a genuine interest in knowledge and a real curiosity about the field will go a long way.

If you are on Twitter, check out #MentoringMonday, which pairs less experienced people with professional mentors. Many, many infosec people are participating. They message each other to mentor each other, answer questions, help find employment, and more. It includes all areas of technology, especially infosec. It's a grassroots movement and it's all free. Use the hashtag #MentoringMonday to post questions and find a mentor.

You don't need a job or an internship just yet—just a connection that could blossom into a mentoring relationship. Follow these important but simple rules for the best results when networking:

- Do your homework about a potential contact, connection, university, school, or employer before you make contact. Be sure to have a general understanding of what they do and why. But don't be a know-it-all. Be open and ready to ask good questions.
- Be considerate of professionals' time and resources. Think about what they can get from you in return for mentoring or helping you.
- Speak and write with proper English. Proofread all your letters, emails, and even texts. Think about how you will be perceived at all times.
- Always stay positive.
- Show your passion for the subject matter.

Here are some places you can start networking:

- *Clubs:* If your school has a computer science club, a hacker club, or something similar, join and be an active member. Invite people you'd like to learn from to come and speak to the club or give the club a tour of their workplace.

- *Volunteering and internships:* These activities give you unique opportunities to learn from and get to know experienced infosec professionals.
- *Professional organizations:* Some of these have student chapters or special events that students can attend.

If you are female and interested in the infosec community, look for a local WoSEC (Women of Security) chapter. This organization was started by Tanya Janca, profiled in this book, with some female friends. It includes women-only learning circles and workshops, helps women build relationships and make connections for jobs, support, and so forth. The goal is to provide support and validation of women in this field, which is dominated by men at the moment. WoSEC currently has thirty-two chapters around the world and is only about two years old. The message is that women can have a great, multidecade career in infosec! Visit https://wearetechwomen.com/wosec-women-of-security/.

- *Social media:* Some social media sites, such as LinkedIn, let you connect with professional organizations as well as people in the field. *However,* just as with all social media, be wary of sharing personal information with people you don't know personally, especially while you are a minor.

Maintain and cultivate professional relationships. Write thank-you notes when professionals take the time to meet with you or share their knowledge and expertise in any form, send updates about your progress and tell them where you decide to go to college, and check in occasionally. If you want to find a good mentor, you need to be a gracious and willing mentee.

Summary

In this chapter, you learned even more about what it's like to work in the infosec field. This chapter discussed the educational requirements of these different areas, including licensing and certifications. You also learned about getting experience in the field before you enter college as well as during the educational process. At this time, you should have a good idea of the educational requirements for your area of interest. You hopefully even contemplated some questions about what kind of educational career path fits your strengths, time requirements, and wallet. Are you starting to picture your career plan? If not, that's okay, as there's still time.

Remember that no matter which area of infosec you pursue, the technologies continually change. You must stay up to date on the latest changes and developments, keep your certifications current, and meet the continuing education requirements. This is very important in the infosec field. Advances in computers are frequent and constant, and it's vitally important that you keep apprised of what's happening in your field. The bottom line is that you need to have a lifelong love of learning to succeed in this field.

"Get a strong foundation and learn from every vehicle you can—online, CBT, etc. Dabble and experiment and *then* decide what you don't like. You will find something that suits your aptitude. You need to find a mentor too—they can guide you and tell you what to try and what to join."—Nadean Tanner, senior manager, Global Technical Education Programs at Puppet (puppet.com)

In chapter 3, we go into a lot more detail about pursing the best educational path. The chapter covers how to find the best value for your education. The chapter includes discussion about financial aid and scholarships. At the end of chapter 3, you should have a much clearer view of the educational landscape and how and where you fit in.

3

Pursuing the Education Path

*W*hen it comes time to start looking at colleges, universities, or postsecondary schools, many high schoolers tend to freeze up at the enormity of the job ahead of them. This chapter will help break down this process for you so it won't seem so daunting.

Yes, finding the right college or learning institution is an important one, and it's a big step toward you achieving your career goals and dreams. The last chapter covered the various educational requirements of the infosec professions, which means you should now be ready to find the right institution of learning. This isn't always just a process of finding the very best school that

It's important to find a postsecondary school that fits your needs and budget. *Getty Images/martine-doucet*

you can afford and can be accepted into, although that may end up being your path. It should also be about finding the right fit so that you can have the best possible experience during your post–high school years.

So here's the truth of it all—attending postsecondary schooling isn't just about getting a degree. It's about learning how to be an adult, managing your life and your responsibilities, being exposed to new experiences, growing as a person, and otherwise moving toward becoming an adult who contributes to society. College—in whatever form it takes for you—offers you an opportunity to actually become an interesting person with perspective on the world and empathy and consideration for people other than yourself, if you let it.

An important component of how successful college will be for you is finding the right fit, the right school that brings out the best in you and challenges you at different levels. I know, no pressure, right? Just as with finding the right profession, your ultimate goal should be to match your personal interests/goals/personality with the college's goals and perspective. For example, small liberal arts colleges have a much different "feel" and philosophy than Big Ten or Pac-12 state schools. And rest assured that all this advice applies even when you're planning on attending community college or another postsecondary school.

Don't worry, though; in addition to these "soft skills," this chapter does dive into the nitty-gritty of finding the best schools, no matter what you want to do. In the infosec field specifically, attending a respected program is important to future success, and we cover that in detail in this chapter.

WHAT IS A GAP YEAR?

Taking a year off between high school and college, often called a gap year, is normal, perfectly acceptable, and almost required in many countries around the world, and it is becoming increasingly acceptable in the United States as well. Even Malia Obama, former president Obama's daughter, did it. Because the cost of college has gone up dramatically, it literally pays for you to know going in what you want to study, and a gap year—well spent—can do lots to help you answer that question.

Some great ways to spend your gap year include joining the Peace Corps or AmeriCorps organizations, enrolling in a mountaineering program or other gap year–styled program, backpacking across Europe or other countries on the cheap (be safe and bring a friend), finding a volunteer organization that furthers a cause you believe in or that complements your career aspirations, joining a Road Scholar program (see www.roadscholar.org), teaching English in another country (see https://www.gooverseas.com/blog/best-countries-for-seniors-to-teach-english-abroad for more information), or working and earning money for college!

Many students will find that they get much more out of college when they have a year to mature and to experience the real world. The American Gap Year Association reports from its alumni surveys that students who take gap years show improved civic engagement, improved college graduation rates, and improved GPAs in college.[1] You can use your gap year to explore and solidify your thoughts and plans about a career in infosec, as well as add impressive experiences to your college application.

Check out https://gapyearassociation.org/ for lots of advice and resources if you're considering a potentially life-altering experience.

Finding the College That's Right for You

Before you look into which schools have degrees in infosec and cybersecurity, it will behoove you to take some time to consider what "type" of school will be best for you. If nothing else, answering questions like the following ones can help you narrow your search and focus on a smaller sampling of choices. Write your answers to these questions down somewhere where you can refer to them often, such as in your Notes app on your phone:

- *Size:* Does the size of the school matter to you? Colleges and universities range from sizes of five hundred or fewer students to twenty-five thousand students.
- *Community location:* Would you prefer to be in a rural area, a small town, a suburban area, or a large city? How important is the location of the school in the larger world to you?
- *Distance from home:* Will you live at home to save money? If not, how far away from home do you want/are you willing to go? Phrase this in terms of hours away or miles away.
- *Housing options:* What kind of housing would you prefer? Dorms, off-campus apartments, and private homes are all common options.

- *Student body:* How would you like the student body to "look"? Think about coed versus all-male and all-female settings, as well as the makeup of minorities, how many students are part-time versus full-time, and the percentage of commuter students. Who will you likely meet there?
- *Academic environment:* Consider which majors are offered and at which levels of degree. Research the student-faculty ratio. Are the classes taught often by actual professors or more often by the teaching assistants? Find out how many internships the school typically provides to students. Are independent study or study abroad programs available in your area of interest?
- *Financial aid availability/cost:* Does the school provide ample opportunities for scholarships, grants, work-study programs, and the like? Does cost play a role in your options (for most people, it does)?
- *Support services:* Investigate the strength of the academic and career placement counseling services of the school.
- *Social activities and athletics:* Does the school offer clubs that you are interested in? Which sports are offered? Are scholarships available?
- *Specialized programs:* Does the school offer honors programs or programs for veterans or students with disabilities or special needs?

> "I've seen people with the very best of educations come out and fail. Education is valuable and can be very substantial, but it's not everything."—Trey Perry, VP of Software Architecture

Not all of these questions are going to be important to you, and that's fine. Be sure to make note of aspects that don't matter so much to you too, such as size or location. You may change your mind as you go to visit colleges, but it's important to make note of where you are to begin with.

CONSIDER THE SCHOOL'S REPUTATION

One factor in choosing a college or certificate program is the school's reputation. This reputation is based on the quality of education previous students have had there. If you go to a school with a healthy reputation in your field, it gives potential employers a place to start when they are considering your credentials and qualifications.

Factors vary depending on which schools offer the program you want, so take these somewhat lightly. Some of the factors affecting reputation generally include:

- *Nonprofit or for-profit:* In general, schools that are nonprofit (or not-for-profit) organizations have better reputations than for-profit schools. In fact, it's best to avoid for-profit schools.
- *Accreditation:* Your program must be accredited by a regional accrediting body to be taken seriously in the professional world. It would be very rare to find an unaccredited college or university with a good reputation.
- *Acceptance rate:* Schools that accept a very high percentage of applicants can have lower reputations than those that accept a smaller percentage. That's because a high acceptance rate can indicate that there isn't much competition for those spaces, or that standards are not as high.
- *Alumni:* What have graduates of the program gone on to do? The college's or department's website can give you an idea of what its graduates are doing.
- *History:* Schools that have been around a long time tend to be doing something right. They also tend to have good alumni networks, which can help you when you're looking for a job or a mentor.
- *Faculty:* Schools with a high percentage of permanent faculty versus adjunct faculty tend to have better reputations. Bear in mind that if you're going to a specialized program or certification program, this may be reversed—these programs are frequently taught by experts who are working in the field.
- *Departments:* A department at one school may have a better reputation than a similar department at a school that's more highly ranked overall. If the department you'll be attending is well known and respected, that could be more important than the overall reputation of the institution itself.

There are a lot of websites that claim to have the "Top 10 Schools for Cybersecurity" or "Best 25 Infosec Programs." It's hard to tell which of those are truly accurate. So where do you begin? *U.S. News & World Report* is a great place to start to find a college or university with a great reputation. Go to www.usnews. com/education to find links to the highest-ranked schools for the undergraduate or graduate degree programs you're interested in.

ROSS ANDERSON: WRITER AND PROFESSOR

Ross Anderson

Ross Anderson, FRS FREng, is professor of security engineering at the University of Cambridge. He received a degree from Cambridge in maths and natural sciences in 1978 and also qualified as a computer hardware engineer. His first job was working in inertial navigation, taking equipment designed for aircraft and repurposing it for use in submarines supporting divers in the North Sea. While writing home computer software in the early 1980s, he became interested in cryptography. In 1986, he began working in the banking industry, helping to secure their cash machines, transfers, and other financial transaction systems. He then helped design and develop prepayment electricity meters, which are now used by four hundred million people in more than one hundred countries. In 1992, he started his PhD in computer science. He so enjoyed doing research that he stayed in the university setting.

He is currently at the Department of Computer Science and Technology, where he is principal investigator at the Cambridge Cybercrime Centre. He is also working on the third edition of his well-known book, Security Engineering.

Can you explain how you became interested in information security as a career path?

I came into the profession via my interest in cryptography, because a friend had been asked to write a program that enabled the partners within a company to share email that other staff couldn't read. I wondered if I could break that program. Eventually, I figured out how to break that cipher and it got me hooked. We developed a better program and this came to the attention of a bank, who hired me. Within a year or two I was designing cryptographic hardware to support payment systems.

What's a "typical" day in your job?

Currently, I am writing the third edition of my *Security Engineering* book. I generally spend about 80 percent of my time doing research and 20 percent teaching, as is normal at research universities. I teach both undergraduate and graduate courses. The undergraduates used to listen to lectures in person; now my lectures are on YouTube and anyone can watch them. The graduate students present a weekly symposium on research papers we send them and then discuss what lessons they have learned.

From the research side, we collect vast amounts of data on spam, phishing, and malware. Our Cambridge Cybercrime Centre makes these available to over one hundred researchers at over fifty universities worldwide, to help them do research on all the bad things that happen online.

How has this field changed over time?

I never get bored because the field changes completely every two years. People haven't stopped being wicked, but their methods evolve over time. Now, people are going online using phones instead of laptops and are increasingly using social media. The *patterns* of fraud, however, really haven't changed much in a decade. Enforcement and opportunity still set the boundaries, regardless of the platform. We're now doing all sorts of measurements to see if the lockdowns worldwide are changing these patterns.

What's the best part of being in this field?

The great thing about being an academic is being paid for doing something that you love, that is your hobby. Basically you just keep doing it, until they carry you out in a body bag!

The nice thing about infosec is that there are so many fields within it. A big service company like Microsoft or Google might have several dozen fraud and abuse teams doing different kinds of work, as well as engineers designing protection mechanisms into the next generation of hardware and software. It's a large field. There are also many specialist firms. Banks and other big companies will always need security experts too, so we're not going to be unemployed anytime soon.

What's the most challenging part of the profession?

With research, the big challenge is finding new, interesting problems to think about.

If you're on the front lines, the core of the job is keeping up with the updates needed to keep things secure. There are constant patches and updates, and to stay on top of it you have to automate most of the work of protecting your systems. You also have to monitor your systems well enough to pick up any signs of an attack.

What is the typical—or best—educational route for students in this field?

Many people get a computer science degree, while some get a specialist master's degree in information security. The ambitious may focus on a research degree, and about half of my former PhD students have ended up doing high-flying jobs in industry and commerce.

What are some things in this profession that are especially challenging right now?

On the technical side, criminals are using Mirai-type botnets of IoT (Internet of Things) devices to launch service-denial attacks. Such botnets are likely to be a problem for some time.

From the research side, the research community is so much larger than it was twenty-five years ago. Rather than having fewer than one hundred people, the top research conferences now have over one thousand. You can't possibly understand all the subdisciplines and know all of the people. You just can't keep up with everything.

Where do you see the field going in the future?

Moore's law (which states that the speed and capability of computers will increase every few years and will be less expensive over time) has finally eased off, and the main platforms are fairly stable. We may be nearing steady state technically. A number of things have to scale up—such as the number of police and investigators hunting down the bad guys.

But we have to put more emphasis on sustainability. At present, software is patched for two to three years on phones and maybe five years on laptops, and then you're expected to buy a new device. This isn't acceptable for durable goods like cars and medical devices. Given that the embedded carbon cost of a car—the amount of carbon we burn to make it—is about as much as the fuel it will burn in its lifetime, we can't afford to start scrapping cars after five years instead of fifteen. So Europe has just passed a law, the Sales of Goods Directive, which means that from January 2022 goods with digital components need maintenance for a period that is a reasonable expectation from the customer. For cars and household appliances this will mean patch support for ten years after the last model leaves the showroom. Software developers will need to support such products for their whole lifetime, which may be twenty years or more. This will bring about big changes in how we do not just software engineering, but safety engineering and security engineering too.

What traits make for a good security analyst?

You need to learn how to do adversarial thinking, as you do when playing chess or go; you need to be able to think of what your opponent can best do against you. When you look at a system, you should be curious about how you could circumvent it. Some people can acquire these skills. Some people are street smart, and some people can never get there.

You also need to have ruthless curiosity and a drive to understand how things really work. You also need to learn enough about the real world to know what the real attacks are likely to be.

What advice do you have for young people considering this career?

You should think about breaking and hacking stuff. Are you good at picking locks? Can you spot loopholes in rules and regulations? If you have that kind of mindset, this might be a good career for you.

How can a young person prepare for this career while in high school?

Learn math and programming to begin with. If you don't like code, try something else.

Any closing comments?

One way in for kids nowadays is gaming. What do you know about people who cheat or who sell game cheats—and how can you stop them? How can you design a game that is very hard to cheat at?

Don't get involved with anything illegal such as DDoS attacks—you can get jail time. You need to know the laws and the limits of experimentation. To get your hands dirty, get engaged in capture-the-flag competitions, whether at school or at a coding club.

AFTER THE RESEARCH, TRUST YOUR GUT

Peter Van Buskirk of the *U.S. News & World Report* puts it best when he says the college that fits you best is one that will do all these things:[2]

- Offers a degree that matches your interests and needs
- Provides a style of instruction that matches the way you like to learn
- Provides a level of academic rigor to match your aptitude and preparation
- Offers a community that feels like home to you
- Values you for what you do well

According to the National Center for Education Statistics (NCES), which is part of the US Department of Education, six years after entering college for an undergraduate degree, only 60 percent of students have graduated. Barely half of those students will graduate from college in their lifetime.

By the same token, it's never been more important to get your degree. College graduates with a bachelor's degree typically earn 66 percent more than those with only a high school diploma and are also far less likely to face unemployment. Also, over the course of a lifetime, the average worker with a bachelor's degree will earn approximately $1 million more than a worker without a postsecondary education.

As you look at the facts and figures, you also need to think about a less quantifiable aspect of choosing a college or university: *fit*. What does that mean? It's hard to describe, but students know it when they feel it. It means finding the school that not only offers the program you want, but also the school that feels right. Many students have no idea what they're looking for in a school until they walk onto the campus for a visit. Suddenly, they'll say to themselves, "This is the one!"

Touring the campus and talking to current students is really important. *Getty Images/SDI Productions*

While you're evaluating a particular institution's offerings with your conscious mind, your unconscious mind is also at work, gathering information about all kinds of things at lightning speed. When it tells your conscious mind what it's decided, we call that a "gut reaction." Pay attention to your gut reactions! There's good information in there.

Hopefully, this section has impressed upon you the importance of finding the right college fit. Take some time to paint a mental picture about the kind of university or school setting that will best meet your needs.

According to the US Department of Education, as many as 32 percent of college students transfer colleges during the course of their educational career. This is to say that the decision you initially make is not set in stone. Do your best to make a good choice, but remember that you can change your mind, your major, and even your campus. Many students do it and go on to have great experiences and earn great degrees.

Honing Your Degree Plan

This section outlines the different approaches you can take to get a degree that will land you your dream job in information security, whether it be as a security analyst, cryptographer, security software developer, cloud security engineer, or something related to all of these.

RELEVANT DEGREE PATHS TO CONSIDER

As you've no doubt learned if you've read this far into the book, the infosec umbrella has many varied, but related, professions within it. No matter which area you want to focus on, having at least an associate's degree (a two-year degree) in computer science or information science is going to give you a leg up during the interviewing process. Consider these points:

- As an associate's degree is a two-year process, it's cheaper and takes less time. Many tech support personnel start their careers with associate's degrees in computer science or information science. Once you are hired, you may be in a position to have your employer pay for you to get your bachelor's degree.

- If you want to enter the workforce as a network engineer, systems analyst, or security analyst, you'll likely need a bachelor's degree (a four-year degree). This could be in computer or information science of course, or maybe even in mathematics, information systems, or engineering.

So, what does the typical computer science degree require of you? Well, as a sampling, the typical computer science student will be required to take the following classes before moving into specifics related to their area of choice:

- Lots of math, including calculus
- Mathematical logic
- Physics
- Electronics
- Statistics and probability

In addition, a typical computer science degree will offer courses on the following subjects:

- Applied computer science
- Coding
- Computer networking
- Microsoft certification
- Operating systems
- Programming languages
- Software engineering
- Database design
- Data logic and management
- Application security, cloud security, network security, and the like

These are just samples of what you will take to gain your degree. Be sure to check the curricula of the schools you're considering attending for more specific information.

STARTING YOUR COLLEGE SEARCH

If you're currently in high school and you are serious about working in the infosec field, start by finding four to five schools in a realistic location (for you) that offer the degree in question. Not every school near you or that you have an initial interest in will probably offer the degree you desire, so narrow your choices accordingly. With that said, consider attending a public university in your resident state, if possible, which will save you lots of money. Private institutions don't typically discount resident student tuition costs.

Be sure you research the basic GPA and SAT or ACT requirements of each school as well.

For those of you applying to associate's degree programs or greater, most advisors recommend that high school students take both the ACT and the SAT tests during their junior year (spring at the latest). (The ACT test is generally considered more weighted in science, so take that into consideration.) You can retake these tests and use your highest score, so be sure to leave time to retake early senior year if needed. You want your best score to be available to all the schools you're applying to by January 1 of your senior year, which will also enable them to be considered with any scholarship applications. (Unless you want to do early decision, which can provide you certain benefits.) Keep in mind these are general timelines—be sure to check the exact deadlines and calendars of the schools to which you're applying! See the section titled "Know the Deadlines" later in this chapter for more information about various deadlines.

Once you have found four to five schools in a realistic location for you that offer the degree you want to pursue, spend some time on their websites studying the requirements for admissions. Most universities will list the average stats for the last class accepted to the program. Important factors weighing on your decision of what schools to apply to should include whether you meet the requirements, your chances of getting in (but shoot high!), tuition costs, availability of scholarships and grants, location, and the school's reputation and licensure/graduation rates.

The order of these characteristics will depend on your grades and test scores, your financial resources, and other personal factors. You of course want to find a university with a good computer science program, and one that also matches your academic rigor and practical needs.

Applying and Getting Admitted

Once you've narrowed down your list of potential schools, of course you'll want to be accepted. First, you need to apply.

There isn't enough room in this book to include everything you need to know about applying to colleges. But here is some useful information to get you started. Remember, every college and university is unique, so be sure to be in touch with their admissions offices so you don't miss any special requirements or deadlines.

Before you go to college, you have to be admitted. *Getty Images/twinsterphoto*

APPLYING TO COLLEGES

It's a good idea to make yourself a "to-do" list while you're a junior in high school. Already a senior? Already graduated? No problem. It's never too late to start.

MAKE THE MOST OF SCHOOL VISITS

If it's at all practical and feasible, you should visit the schools you're considering. To get a real feel for any college or school, you need to walk around the campus and buildings, spend some time in the common areas where students hang out, and sit in on a few classes. You can also sign up for campus tours, which are typically given by current students. This is another good way to see the school and ask questions of someone who knows. Be sure to visit the specific school/building that covers your possible major as well. The website and brochures won't be able to convey that intangible feeling you'll get from a visit.

In addition to the questions listed near the beginning of this chapter titled "Finding the College That's Right for You," consider these questions as well. Make a list of questions that are important to you before you visit, such as:

- What is the makeup of the current freshman class? Is the campus diverse?
- What is the meal plan like? What are the food options?
- Where do most of the students hang out between classes? (Be sure to visit this area.)
- How long does it take to walk from one end of the campus to the other?
- What types of transportation are available for students? Does campus security provide escorts to cars, dorms, and so forth at night?

In order to be ready for your visit and make the most of it, consider these tips and words of advice.

Before you go:

- Make a list of questions.
- Be sure to do some research. At the least, spend some time on the college website. Make sure your questions aren't addressed adequately there first.
- Arrange to meet with a professor in your area of interest or to visit the specific school.

- Be prepared to answer questions about yourself and why you are interested in this school.
- Dress in neat, clean, and casual clothes. Avoid overly wrinkled clothing or anything with stains.

When you're there:

- Listen and take notes.
- Don't interrupt.
- Be positive and energetic.
- Make eye contact when someone speaks directly to you.
- Ask questions.
- Thank people for their time.

Finally, be sure to send thank-you notes or emails after the visit is over. Remind the recipient when you visited the campus and thank them for their time.

STANDARDIZED TESTS

Many colleges and universities require scores from standardized tests that are supposed to measure your readiness for college and ability to succeed. There is debate about how accurate these tests are, so some institutions don't ask for them anymore. But most do, so you should expect to take them.

Undergraduate-Level Tests

To apply to an undergraduate program, students generally take either the SAT or the ACT. Both cover reading, writing, and math. Both have optional essays. Both are accepted by colleges and universities. Both take nearly the same amount of time to complete. If one test is preferred over another by schools, it's usually more about where you live than about the test.[3]

- *SAT:* Offered by the College Board (CollegeBoard.org). There are twenty SAT subject tests that you can take to show knowledge of special areas, such as math 1 and math 2, biology (ecological or molecular), chemistry, and physics, as well as US or world history and numerous languages.

- *ACT:* Offered by ACT.org. There aren't any subject tests available with the ACT. Questions are a little easier on the ACT, but you don't have as much time to answer them.

Ultimately, which test you take comes down to personal preference. Many students choose to take both exams.

Graduate-Level Tests

- *GRE (Graduate Record Exam):* Published by ETS (Educational Testing Service). The GRE is the most widely used admission test for graduate and professional schools. It covers verbal and quantitative reasoning and analytical writing. The test results are considered along with your undergraduate record for admissions decisions to most graduate programs.
- *GRE subject tests:* Some graduate programs also want to see scores from subject tests. GRE subject tests are offered in biology, chemistry, literature in English, mathematics, physics, and psychology.
- *MCAT (Medical College Admission Test):* Administered by the Association of American Medical Colleges (AAMC). MCAT is the standardized test for admission to medical school programs in allopathic, osteopathic, podiatric, or veterinary medicine (some veterinary programs accept the GRE instead).
- *LSAT (Law School Admission Test):* Administered by the Law School Admission Council seven times a year. This test for prospective law school candidates is the only test accepted for admission purposes by all ABA-accredited law schools and Canadian common-law law schools.

KNOW THE DEADLINES

- *Early decision (ED)* deadlines are usually in November, with acceptance decisions announced in December. Note that if you apply for ED admission and are accepted, that decision is binding, so only apply for ED if you know exactly which school you want to go to and are ready to commit.
- *ED II* is a second round of early decision admissions. Not every school that does ED will also have an ED II. For those that do, deadlines are usually in January with decisions announced in February.

- *Regular decision* deadlines can be as early as January 1 but can go later. Decision announcements usually come out between mid-March and early April.
- *Rolling admission* is used by some schools. Applications are accepted at any time, and decisions are announced on a regular schedule. Once the incoming class is full, admissions for that year will close.

THE COMMON APP

The Common Application form is a single, detailed application form that is accepted by more than nine hundred colleges and universities in the United States. Instead of filling out a different application form for every school you want to apply to, you fill out one form and have it sent to all the schools you're interested in. The Common App itself is free, and most schools don't charge for submitting it.

If you don't want to use the Common App for some reason, most colleges will also let you apply with a form on their website. There are a few institutions that only want you to apply through their sites and other highly regarded institutions that only accept the Common App. Be sure you know what the schools you're interested in prefer.

The Common App's website (www.commonapp.org) has a lot of useful information, including tips for first-time applicants and for transfer students.

ESSAYS

Part of any college application is a written essay, sometimes even two or three. Some colleges provide writing prompts they want you to address. The Common App has numerous prompts that you can choose from. Here are some issues to consider when writing your essays:

- *Topic:* Choose something that has some meaning for you and that you can speak to in a personal way. This is your chance to show the college or university who you are as an individual. It doesn't have to be about an achievement or success, and it shouldn't be your whole life story. Maybe a topic relates to a time you learned something or had an insight into yourself.

- *Timing:* Start working on your essays the summer before senior year, if possible. You won't have a lot of other homework in your way, and you'll have time to prepare thoughtful comments and polish your final essay.
- *Length:* Aim for between 250 and 650 words. The Common App leans toward the long end of that range, while individual colleges may lean toward the shorter end.
- *Writing:* Use straightforward language. Don't turn in your first draft—work on your essay, improve it as you go. Ask someone else to read it and tell you what they think. Ask your English teacher to look at it and make suggestions. Do NOT let someone else write any portion of your essay. It needs to be *your* ideas and *your* writing in order to represent *you.*
- *Proofing:* Make sure your essay doesn't have any obvious errors. Run spell check, but don't trust it to find everything (spell checkers are notorious for introducing weird errors). Have someone you trust read it over for you and note spelling, grammar, and other mistakes. Nobody can proofread their own work and find every mistake—what you'll see is what you expect to see. Even professional editors need other people to proofread their writing! So don't be embarrassed to ask for help.

LETTERS OF RECOMMENDATION

Most college applications ask for letters of recommendation from people who know you well and can speak to what you're like as a student and as a person. How many you need varies from school to school, so check with the admissions office website to see what they want. Some schools don't want any!

Whom Should You Ask for a Letter?

Some schools will tell you pretty specifically whom they want to hear from. Others leave it up to you. Choose people who know you and think well of you, such as:

- One or two teachers of your best academic subjects (English, math, science, social studies, etc.)
- Teacher of your best elective subject (art, music, media, etc.)
- Advisor for a club you're active in

- School counselor
- School principal (but only if you've taken a class with them or they know you individually as a student)
- Community member you've worked with, such as a scout leader, volunteer group leader, or religious leader
- Boss at a job you've held

When Should You Ask for a Letter?

Don't wait until applications are due. Give people plenty of time to prepare a good recommendation letter for you. If possible, ask for these letters in late spring or early summer of your junior year.

Submitting Your Letters of Recommendation

Technically, you're not supposed to read your recommendation letters. That lets recommenders speak more freely about you. Some may show you the letter anyway, but that's up to them. Don't ask to see it!

Recommenders can submit their letters electronically either directly to the institutions you're applying to or through the Common App. Your job is to be sure they know the submission deadlines well in advance so they can send in the letters on time.

ADMISSIONS REQUIREMENTS

Each college or university will have its own admissions requirements. In addition, the specific program or major you want to go into may have admissions requirements of its own, in addition to the institution's requirements.

It's your responsibility to go to each institution's website and be sure you know and understand their requirements. That includes checking out each department site, too, to find any special prerequisites or other things that they're looking for.

THE MOST PERSONAL OF PERSONAL STATEMENTS

The *personal statement* you include with your application to college is extremely important, especially when your GPA and SAT/ACT scores are on the border of what is typically accepted. Write something that is thoughtful and conveys your understanding of the IT profession, as well as your desire to work in the information technology world. Why are you uniquely qualified? Why are you a good fit for this university and program or these types of students? These essays should be highly personal (the "personal" in personal statement). Will the admissions professionals who read it, along with hundreds of others, come away with a snapshot of who you really are and what you are passionate about?

Look online for some examples of good ones, which will give you a feel for what works. Be sure to check your specific school for length guidelines, format requirements, and any other guidelines they expect you to follow. Most important, make sure your passion for your potential career comes through—although make sure it is also genuine.

And of course, be sure to proofread it several times and ask a professional (such as your school writing center or your local library services) to proofread it as well.

What's It Going to Cost You?

So, the bottom line—what will your education end up costing you? Of course that depends on many factors, including the type and length of degree, where you attend (in-state or not, private or public institution), how much in scholarships or financial aid you're able to obtain, your family or personal income, and many other factors.

"College may seem expensive. But the truth is that most students pay less than their college's sticker price, or published price, thanks to financial aid. So instead of looking at the published price, concentrate on your net price—the real price you'll pay for a college. . . . Your net price is a college's sticker price for tuition and fees minus the grants, scholarships, and education tax benefits you receive. The net price you pay for a particular college is specific to you because it's based on your personal circumstances and the college's financial aid policies."—BigFuture

Table 3.1. Average Estimated Full-Time Undergraduate Budgets (Enrollment Weighted) by Sector, 2018–2019

Sector	Tuition and Fees	Room and Board	Books and Supplies	Transportation	Other Expenses	Total
Private Nonprofit Four-Year On-Campus	$35,830	$12,680	$1,240	$1,050	$1,700	$52,500
Public Four-Year Out-of-State On-Campus	$26,290	$11,140	$1,240	$1,160	$2,120	$41,950
Public Four-Year In-State On-Campus	$10,230	$11,140	$1,240	$1,160	$2,120	$25,890
Public Two-Year In-District Commuter	$3,660	$8,660	$1,440	$1,800	$2,370	$17,930

Note: Expense categories are based on institutional budgets for students as reported in the College Board's Annual Survey of Colleges. Figures for tuition and fees and room and board mirror those reported in the table above. Other expense categories are the average amounts allotted in determining the total cost of attendance and do not necessarily reflect actual student expenditures.

Source: Jennifer Ma, Sandy Baum, Matea Pender, and C. J. Libassi, *Trends in College Pricing 2019* (New York: College Board, 2019), https://research.collegeboard.org/trends/college-pricing.

The College Board (see www.collegeboard.org) tracks and summarizes financial data from colleges and universities all over the United States. A sample of the most recent data is shown in table 3.1. It represents the state of things for the 2018–2019 academic year. (It's worth noting that these numbers represent a 2.5 percent increase over 2017–2018 costs before adjusting for inflation.) Costs shown are for one year.

Keep in mind these are averages and reflect the published prices, not the net prices. As an example of net cost, in 2018–2019, full-time in-state students at public four-year colleges must cover an average of about $14,900 in tuition and fees and room and board after grant aid and tax benefits, in addition to paying for books and supplies and other living expenses.[4]

If you read more specific data about a particular university or find averages in your particular area of interest, you should assume those numbers are closer to reality than these averages, as they are more specific. This data helps to show you the ballpark figures.

Generally speaking, there is about a 3 percent annual increase in tuition and associated costs to attend college. In other words, if you are expecting to attend college two years after this data was collected, you need to add approximately 6 percent to these numbers. Keep in mind this assumes no financial aid or scholarships of any kind (so it's not the net cost).

This chapter also covers finding the most affordable path to get the degree you want. You'll also learn how to prime the pumps and get as much money for college as you can.

Financial Aid and Student Loans

Finding the money to attend college, whether it is two or four years, an online program, or a vocational career college, can seem overwhelming. But you can do it if you have a plan before you actually start applying to college.

NOT ALL FINANCIAL AID IS CREATED EQUAL

Educational institutions tend to define financial aid as any scholarship, grant, loan, or paid employment that assists students to pay their college expenses. Notice that "financial aid" covers both *money you have to pay back* and *money you don't have to pay back*. That's a big difference!

Do Not Have to Be Repaid

- Scholarships
- Grants
- Work-study

Have to Be Repaid *with Interest*

- Federal government loans
- Private loans
- Institutional loans

If you get into your top-choice university, don't let the sticker cost turn you away. Financial aid can come from many different sources, and it's available to cover all different kinds of costs you'll encounter during your years in college, including tuition, fees, books, housing, and food.

The good news is that universities more often offer incentives or tuition discount aid to encourage students to attend. The market is often more competitive in the favor of the student, and colleges and universities are responding by offering more generous aid packages to a wider range of students than they used to. Here are some basic tips and pointers about the financial aid process:

- You apply for financial aid during your senior year. You must fill out the FAFSA (Free Application for Federal Student Aid) form, which can be filed starting October 1 of your senior year until June of the year you graduate.[5] Because the amount of available aid is limited, it's best to apply as soon as you possibly can. See fafsa.gov to get started.
- Be sure to compare and contrast deals you get at different schools. There is room to negotiate with universities. The first offer for aid may not be the best you'll get.

Paying for college can take a creative mix of grants, scholarships, and loans, but you can find your way with some help! *Getty Images/zimmytws*

- Wait until you receive all offers from your top schools, and then use this information to negotiate with your top choice to see if they will match or beat the best aid package you received.
- To be eligible to keep and maintain your financial aid package, you must meet certain grade/GPA requirements. Be sure you are very clear on these academic expectations and keep up with them.
- You must reapply for federal aid every year.

Watch out for scholarship scams! You should never be asked to pay to submit the FAFSA form ("free" is in its name) or be required to pay a lot to find appropriate aid and scholarships. These are free services. If an organization promises you you'll get aid or that you have to "act now or miss out," these are both warning signs of a less reputable organization.

Also, be careful with your personal information to avoid identity theft as well. Simple things like closing and exiting your browser after visiting sites where you entered personal information (like fafsa.gov) goes a long way. Don't share your student aid ID number with anyone either.

It's important to understand the different forms of financial aid that are available to you. That way, you'll know how to apply for different kinds and get the best financial aid package that fits your needs and strengths. The two main categories that financial aid falls under is *gift aid*, which doesn't have to be repaid, and *self-help aid*, which are either loans that must be repaid or work-study funds that are earned. The next sections cover the various types of financial aid that fit in one of these areas.

GRANTS

Grants typically are awarded to students who have financial needs but can also be used in the areas of athletics, academics, demographics, veteran support, and special talents. They do not have to be paid back. Grants can come from federal agencies, state agencies, specific universities, and private organizations. Most federal and state grants are based on financial need. Examples of grants are the Pell Grant and the SMART Grant.

SCHOLARSHIPS

Scholarships are merit-based aid that does not have to be paid back. They are typically awarded based on academic excellence or some other special talent, such as music or art. Scholarships also fall under the areas of athletic based, minority based, aid for women, and so forth. These are typically not awarded by federal or state governments but instead come from the specific school you applied to as well as private and nonprofit organizations.

Be sure to reach out directly to the financial aid officers of the schools you want to attend. These people are great contacts who can lead you to many more sources of scholarships and financial aid. Visit http://www.gocollege. com/financial-aid/scholarships/types/ for lots more information about how scholarships in general work.

LOANS

Many types of loans are available especially to students to pay for their post-secondary education. However, the important thing to remember here is that loans *must be paid back, with interest*. Be sure you understand the interest rate you will be charged. This is the extra cost of borrowing the money and is usually a percentage of the amount you borrow. Is this fixed or will it change over time? Are the loan and interest deferred until you graduate (meaning you don't have to begin paying it off until after you graduate)? Is the loan subsidized (meaning the federal government pays the interest until you graduate)? These are all points you need to be clear about before you sign on the dotted line.

There are many types of loans offered to students, including need-based loans, non-need-based loans, state loans, and private loans. Two very reputable federal loans are the Perkins Loan and the Direct Stafford Loan. For more information about student loans, start at https://bigfuture.collegeboard.org/pay-for-college/loans/types-of-college-loans.

FEDERAL WORK-STUDY

The U.S. Federal Work-Study Program provides part-time jobs for undergraduate and graduate students with financial need so they can earn money to pay for educational expenses. The focus of such work is on community service work and work related to a student's course of study. Not all schools participate in this program, so be sure to check with the school financial aid office if this is something you are counting on. The sooner you apply, the more likely you will get the job you desire and be able to benefit from the program, as funds are limited. See https://studentaid.ed.gov/sa/types/work-study for more information about this opportunity.

FINANCIAL AID TIPS

- Some colleges/universities will offer tuition discounts to encourage students to attend—so tuition costs can be lower than they look at first.
- Apply for financial aid during your senior year of high school. The sooner you apply, the better your chances.
- Compare offers from different schools—one school may be able to match or improve on another school's financial aid offer.
- Keep your grades up—a good GPA helps a lot when it comes to merit scholarships and grants.
- You have to reapply for financial aid every year, so you'll be filling out that FAFSA form again!
- Look for ways that loans may be deferred or forgiven—service commitment programs are a way to use service to pay back loans.

While You're in College

Once you're in an undergrad program, of course you'll take all the classes required by your major. This will be time consuming and a lot of hard work, as it should be. But there's more to your college experience than that!

One of the great advantages of college is that it's so much more than just training for a particular career. It's your opportunity to become a broader, deeper person. Use your electives to take courses far outside your major. Join clubs, intramural teams, improv groups—whatever catches your interest. Take a foreign language. The broader your worldview is, the more interesting you are as a person. And the more appealing you are to employers in the future!

> "It has always seemed strange to me that in our endless discussions about education so little stress is laid on the pleasure of becoming an educated person, the enormous interest it adds to life. To be able to be caught up into the world of thought—that is to be educated."—Edith Hamilton

WORKING WHILE YOU LEARN

Your classes won't always convey what it's like to do the work in real life, especially in the cutting-edge world of cybersecurity. If you have the opportunity, consider some of these ways to learn and work at the same time.

Cooperative Education Programs

Cooperative education (co-op) programs are a structured way to alternate classroom instruction with on-the-job experience. There are co-op programs for all kinds of jobs, especially engineering programs. Co-op programs are run by the educational institution in partnership with several employers. Students usually alternate semesters in school with semesters at work.

A co-op program is not an internship. Students in co-op jobs typically work forty hours a week during their work semesters and are paid a regular salary. Participating in a co-op program means it will take longer to graduate, but you come out of school with a lot of legitimate work experience.

Be sure the college you attend is truly committed to its co-op program. Some are deeply committed to the idea of co-ops as integral to education, but others treat it more like an add-on program. Also, the company you co-op with is not obliged to hire you at the end of the program. But they can still be an excellent source of good references for you in your job search.

Internships

Internships are another way to gain work experience while you're in school. Internships are offered by employers and usually last one semester or one summer. You could work part-time or full-time, but you're usually paid in experience and college credit rather than money. There are paid internships in some fields, but they aren't common.

Making High School Count

If you are still in high school or middle school, there are many things you can do now to nurture your interest in informational technology and cybersecurity and set yourself up for success. Consider these tips for your remaining years:

- Work on listening well and speaking and communicating clearly. Work on writing clearly and effectively.
- Learn how to learn. This means keeping an open mind, asking questions, asking for help when you need it, taking good notes, and doing your homework.
- Plan a daily homework schedule and keep up with it. Have a consistent, quiet place to study.
- Talk about your career interests with friends, family, and counselors. They may have connections to people in your community whom you can shadow or who will mentor you.
- Try new interests or activities, especially during your first two years of high school.
- Be involved in extracurricular activities that truly interest you and say something about who you are and want to be.

Remember to take care of yourself and to enjoy the journey to adulthood! *Getty Images/AntonioGuillem*

Kids are under so much pressure these days to "do it all," but you should think about working smarter rather than harder. If you are involved in things you enjoy, your educational load won't seem like such a burden. Be sure to take time for self-care, such as sleep, unscheduled downtime, and other activities that you find fun and energizing. See chapter 4 for more ways to relieve and avoid stress.

Summary

This chapter dove right in and talked about all the aspects of college and post-secondary schooling that you'll want to consider as you move forward. Remember that finding the right fit is especially important, as it increases the chances that you'll stay in school and finish your degree or program, as well as have an amazing experience while you're there.

In this chapter, you learned a little about what a typical computer science degree will require of you. You also learned about how to get the best education for the best deal. You learned a little about scholarships and financial aid, how the SAT and ACT tests work, how to write a unique personal statement that eloquently expresses your passions, and how to do your best at essays and other application requirements.

Use this chapter as a jumping-off point to dig deeper into your particular area of interest. Some tidbits of wisdom to leave you with:

- If you need to, take the SAT and ACT tests early in your junior year so you have time to take them again. Most schools automatically accept the highest scores (but be sure to check your specific schools' policies).
- Don't underestimate how important school visits are, especially in the pursuit of finding the right academic fit. Come prepared to ask questions not addressed on the school website or in the literature.
- Your personal statements/essays are very important pieces of your application that can set you apart from others. Take the time and energy needed to make them unique and compelling.
- Don't assume you can't afford a school based on the "sticker price." Many schools offer great scholarships and aid to qualified students. It doesn't hurt to apply. This advice especially applies to minorities, veterans, and students with disabilities.
- Don't lose sight of the fact that it's important pursue a career that you enjoy, are good at, and are passionate about! You'll be a happier person if you do so.

"If anyone ever tells you can't or you don't belong in infosec or you must take all these expensive certifications or otherwise tries to put up other blockades, just ignore them and make a note to yourself that they give bad advice. There are gatekeepers out there, and you shouldn't follow advice from them. Sometimes it's jealousy they did all that stuff and they think you must too. But you can safely avoid and ignore their advice. Find the people saying, 'You can do it—here's how. Follow me—this way.'"—Tanya Janca, founder, security trainer, and coach of SheHacks Purple.dev, specializing in training others in software and cloud security

At this point, your career goals and aspirations should be gelling. At the least, you should have a plan for finding out more information. And don't forget about networking, which was covered in more detail in chapter 2. Remember to do the research about the school or degree program before you reach out and especially before you visit. Faculty and staff find students who ask challenging questions much more impressive than those who ask questions that can be answered by spending ten minutes on the school website.

In chapter 4, we go into detail about the next steps—including writing a resume and cover letter, interviewing well, and follow-up communications. This is information you can use to secure internships, volunteer positions, summer jobs, and more. It's not just for college grads. In fact, the sooner you can hone these communication skills, the better off you'll be in the professional world, regardless of your job.

4

Writing Your Resume and Interviewing

No matter which area of infosec you aspire to work in, having a well-written resume and impeccable interviewing skills will help you reach your ultimate goals. This chapter provides some helpful tips and advice to build the best resume and cover letter, how to interview well with all your prospective employers, and how to communicate effectively and professionally at all times. All the advice in this chapter isn't just for people entering the workforce full-time either. It can help you score that internship or summer job or help you give a great college interview to impress the admissions office.

After we talk about writing your resume, the chapter discusses important interviewing skills that you can build and develop over time. The chapter also has some tips for using social media to your benefit, as well as how to deal successfully with stress, which is an inevitable by-product of a busy life. Let's dive in!

Writing Your Resume

If you're a teen writing a resume for your first job, you likely don't have a lot of work experience under your belt yet. Because of this limited work experience, you need to include classes and coursework that are related to the job you're seeking, as well as any school activities and volunteer experience you have. While you are writing your resume, you may discover some talents and recall some activities you did that you forgot about, which are still important to add. Think about volunteer work, side jobs you've held (help desk work, tutoring, troubleshooting your neighbor's computer, etc.), and the like. A good approach at this point in your career is to build a functional-type resume, which focuses on your abilities rather than work experience, and it's discussed in detail next.

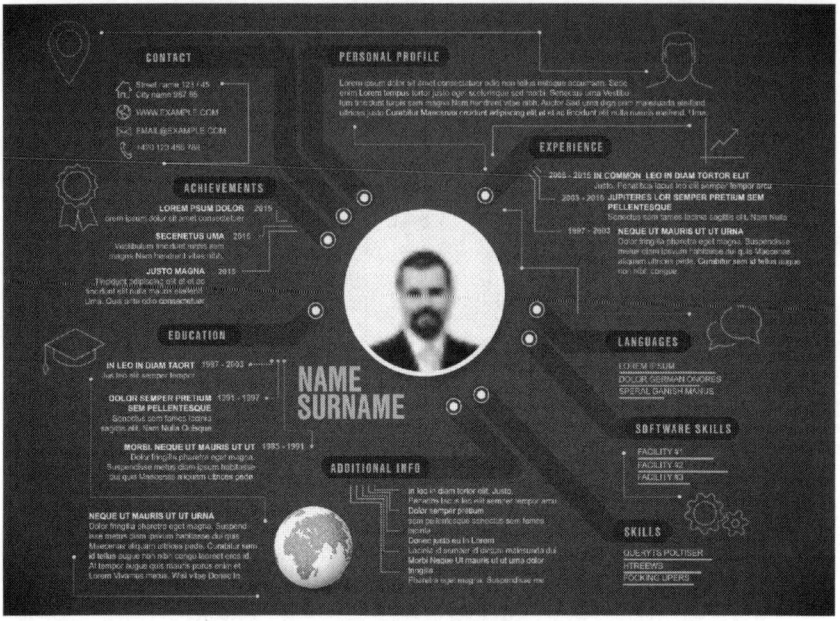

Just because you're in a technical field, it doesn't mean your resume has to be boring looking. Just make sure it's functional and easy to read as well. *Getty Images/orsonsurf*

PARTS OF A RESUME

As mentioned, the functional resume is the best approach when you don't have a lot of pertinent work experience, as it is written to highlight your abilities rather than the experience. The other, perhaps more common, type of resume is called the chronological resume, and it lists a person's accomplishments in chronological order, most recent jobs listed first. This section breaks down and discusses the functional resume in greater detail.

Here are the essential parts of your resume, listed from the top down:

- *Heading*—This should include your name, address, and contact information, including phone, email, and website if you have one. This information is typically centered on the page.
- *Objective*—This is one sentence that tells that specific employer what kind of position you are seeking. These should be modified to be specific to each potential employer.

- *Education*—Always list your most recent school or program first. Include date of completion (or expected date of graduation), degree or certificate earned, and the institution's name and address. Include workshops, seminars, and related classes here as well.
- *Skills*—Skills include computer literacy, leadership skills, organizational skills, or time-management skills. Be specific in this area when possible, and tie them to working with computers when it's appropriate.
- *Activities*—These can be related to skills. Perhaps an activity listed here led to you developing a skill listed above. This section can be combined with the skills section, but it's often helpful to break these apart if you have enough substantive things to say in both areas. Examples include camps, sports teams, leadership roles, community service work, clubs, and organizations, as well as anytime you worked on computers.
- *Experience*—If you don't have any actual work experience that's relevant, you may consider skipping this section. However, you can list summer, part-time, and volunteer jobs you've held, again focusing on work related to IT and infosec.
- *Interests*—This section is optional, but it's a chance to include special talents and interests. Keep it short, factual, and specific.
- *Languages*—List all the programming languages you've used, as well as relevant software, antivirus ware, and so on.
- *References*—It's best to say that references are available on request. If you do list actual contacts, list no more than three and make sure you inform your contacts that they may be contacted.

The skills, experience, interests, and languages entries can be creatively combined or developed to maximize your abilities and experience. These are not set-in-stone sections that every resume must have.

> "Whatever it is you are good at, use that as your strength and see how it fits into the industry if that's what you want to do. For instance, I am good at writing and talking about the industry and that found me a place in infosec. You can find a way to make it work with your unique talents. There is such a wide breadth of options and you can use your talents to find a way in."—Ben Malisow, CISSP, CISM, CCSP, and Security+

Even just helping fellow students with software and hardware problems can be a good thing to mention as experience. *Getty Images/kate_sept2004*

If you're still not seeing the big picture here, it's helpful to look at student and part-time resume examples online to see how others have approached this process. Search for "functional resume examples" to get a look at some examples.

RESUME-WRITING TIPS

Regardless of your situation and why you're writing the resume, there are some basic tips and techniques you should use:

- Keep it short and simple. This includes using a simple, standard font and format. Using one of the resume templates provided by your word processor software can be a great way to start.
- Use simple language. Keep it to one page.
- Highlight your academic achievements, such as a high GPA (above 3.5) or academic awards. If you have taken classes related to the job you're interviewing for, list those briefly as well.

- Emphasize your extracurricular activities, internships, and the like. These could include camps, clubs, sports, tutoring, or volunteer work. Use these activities to show your skills, interests, and abilities.
- Use action verbs, such as "led," "created," "taught," "ran," "developed."
- Be specific and give examples.
- Always be honest.
- Include leadership roles and experience.
- Edit and proofread at least twice and have someone else do the same. Ask a professional (such as your school writing center or your local library services) to proofread it for you also. Don't forget to run spell check.
- Include a cover letter if necessary (discussed next).

THE COVER LETTER/EMAIL INTRO LETTER

Every resume you send out through the standard mail should include a cover letter. If you are sending your resume via email or another electronic format, be sure to provide introductory text that serves the place of the traditional cover letter. This can be the most important part of your job search because it's often the first thing that potential employers read. By including this, you're showing the employers that you took the time to learn about their organization and address them personally. This goes a long way to show that you're interested in the position.

Be sure to call the company or verify on the website the name and title of the person to whom you should address the letter/email. This letter/email should be brief. Introduce yourself and begin with a statement that will grab the person's attention. Keep in mind that they will potentially be receiving hundreds of resumes and cover letters for an open position.

As with the resume, you have a little room to be creative, but the cover letter should contain the following parts in order from top to bottom:

- Your name, address, phone number, and email address
- Today's date
- The recipient's name, title, company name, and company address
- Salutation
- Body (usually one to three paragraphs)
- Closing (your signature and name)

An email would not include the address or date. The body portion should mention how you heard about the position, something extra about you that will interest the potential employer, practical skills you can bring to the position, and past experience related to the job. You should apply the facts outlined in your resume to the job to which you're applying.

Your cover letter/email should be as short as possible while still conveying a sense of who you are and why you want this particular job or to work for this particular company. Do your research into the company and include some details about the company in your letter/email—this demonstrates that you cared enough to take the time to learn something about them.

> Always try to find out the name and title of the person who will be handling your application. This is usually listed in the job posting, but if not, taking the time to track it down yourself on the company website can pay off. Think about it. Imagine that you read nine cover letters addressed to "Dear sir or madam" and then you get a tenth addressed to you using your name. Wouldn't that one catch your attention?

Here are a few more things to keep in mind when writing your cover letter/ email:

- As with the resume, proofread it many times and have others proofread it too. You don't want them to discard a great letter you worked so hard on just because you forgot to finish a sentence or made some clumsy mistake.
- Use "Mr." for male names and "Ms." for female names in your salutation. If you can't figure out the sex of the person who will be handling your application from their name, just use their full name ("Dear Jamie Smith").
- As much as possible, connect the specific qualifications the company is looking for with your skill set and experience. The closer you can make yourself resemble the ideal candidate described in the ad, the more likely they will call you for an interview.
- If you're lacking one or more qualifications the employer is looking for, just ignore it. Don't call attention to it by pointing it out or making an excuse. Leave it up to them to decide how important it is and whether

they still want to call you in for an interview. Your other skills and experience may more than make up for any one lack.

- You want to seem eager, competent, helpful, and dependable. Think about things from the employer's point of view. Focus your letter much more on what *you* can do for *them* than on what they can do for you. Be someone that *you* would want to hire.
- For more advice on cover letters, check out the free guide by Resume Genius at https://resumegenius.com/cover-letters-the-how-to-guide.

RESUMES, COVER LETTERS, AND ONLINE JOB APPLICATIONS

Resumes and cover letters are holdovers from the era before the internet—from before personal computers even. They were designed to be typed on paper and delivered through the mail. Obviously nowadays much of the job application process has moved online. Nevertheless, the essential concepts communicated by the resume and cover letter haven't changed. Most employers who accept online applications either ask that you email or upload your resume.

Those who ask you to email your resume will specify which document formats they accept. The Adobe Acrobat PDF format is often preferred, because many programs can display a PDF (including web browsers), and documents in this format are mostly uneditable—that is, they can't easily be changed. Most word processing programs have an option under the Save command that allows you to save your work as a PDF.

In these cases, you attach the resume to your email, and your email itself becomes the "cover letter." The same principles of the cover letter discussed in this section apply to this email, except you skip the addresses and date at the top and begin directly with the salutation. You'll also need to pay particular attention to the subject line of your email. Be sure that it is specific to the position you are applying for.

In all cases, it's really important to follow the employer's instructions on how to submit your cover letter and resume. Some employers direct you to a section on their website where you can upload your resume. In these cases, it may not be obvious where your cover letter content should go. Look for a textbox labeled something like "Personal Statement" or "Additional Information." Those are good places to add whatever you would normally write in a cover letter. If there doesn't seem to be anywhere like that, see if there is an email link to the hiring manager or whoever will be reading your resume. Go ahead and send your "cover email" to this address, mentioning that you have uploaded your resume (again omitting the addresses and date at the top of your cover letter). Try to use the person's name if it's been given.

LINKING-IN WITH IMPACT

As well as your paper or electronic resume, creating a LinkedIn profile is a good way to highlight your experience and promote yourself, as well as to network. Joining professional organizations or connecting with other people in your desired field are good ways to keep abreast of changes and trends and work opportunities.

The key elements of a LinkedIn profile are your photo, your headline, and your profile summary. These are the most revealing parts of the profile and the ones employers and connections will base their impression of you on.

The photo should be carefully chosen. Remember that LinkedIn is not Facebook or Instagram; it is not the place to share a photo of you acting too casually on vacation or at a party. According to Joshua Waldman, author of *Job Searching with Social Media for Dummies*, the choice of photo should be taken seriously and be done right. His tips:[1]

- Choose a photo in which you have a nice smile.
- Dress in professional clothing.
- Ensure the background of the photo is pleasing to the eye. According to Waldman, some colors—like green and blue—convey a feeling of trust and stability.
- Remember, it's not a mug shot. You can be creative with the angle of your photo rather than stare directly into the camera.
- Use your photo to convey some aspect of your personality.
- Focus on your face. Remember visitors to your profile will see only a small thumbnail image, so be sure your face takes up most of it.

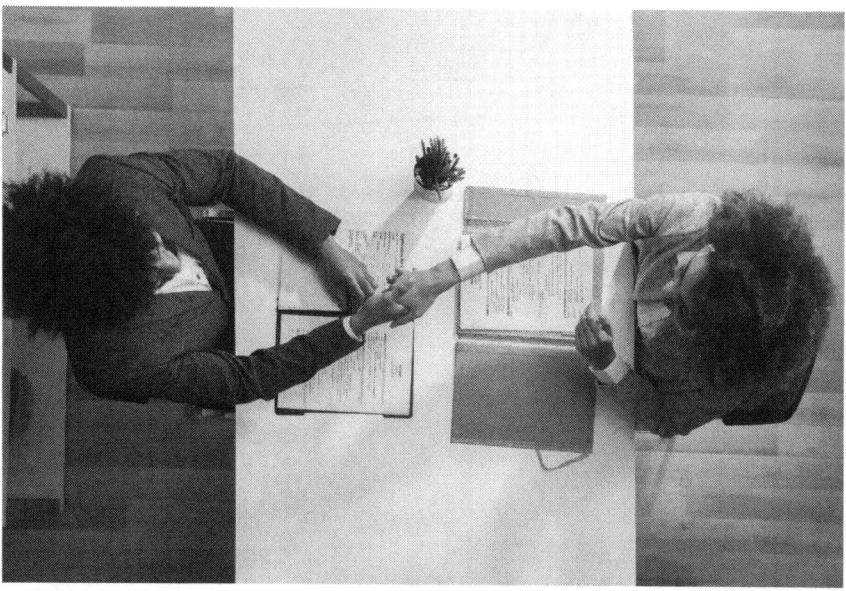

Being prepared will help you feel less stressed during a job interview. *Getty Images/AndreyPopov*

Interviewing Skills

Sooner or later, your job search will result in what you've been hoping for (or, perhaps, dreading): a phone call or email requesting that you appear for an interview. When you get that call or email, it means they're interested in you and are considering hiring you.

"Be sure to balance your focus on the big picture with intense attention to detail and communication. It's easy in this space to wave your hands and say we should do x or y differently, but you need to understand why they did that and understand the details of making those changes before you are actually any good. Deep technical understanding must be balanced with higher-level idea of what to change and why you will or won't be effective."—Adam Shostack, noted expert on threat modeling, as well as a technologist, author, and game designer

WHAT ARE EMPLOYERS LOOKING FOR?

When companies want to hire new employees, they list job descriptions on job hunting websites. These are a fantastic resource for you long before you're ready to actually apply for a job. You can read real job descriptions for real jobs and see what qualifications and experience are needed for the kinds of job you're interested in. You can see what sort of tasks you'll be carrying out in different kinds of jobs. You'll also get a good idea of the range of salaries and benefits that go with different levels of experience.

Pay attention to the "Required Qualifications," of course, but also pay attention to the "Desired Qualifications"—these are the ones you don't *have* to have, but if you have them, you'll have an edge over other potential applicants.

Here are a few to get you started:

www.monster.com

www.indeed.com

www.ziprecruiter.com

www.glassdoor.com

www.simplyhired.com

The best way to avoid nerves and keep calm when you're interviewing is to be prepared. It's okay to feel scared, but keep it in perspective. It's likely that you'll receive many more rejections in your professional life than acceptances, as we all do. However, you only need one "yes" to start out. Think of the interviewing process as a learning experience.

Personal contacts can make the difference! Don't be afraid to contact other professionals you know. Personal connections can be a great way to find jobs and internship opportunities. Your high school teachers, your coaches and mentors, and your friends' parents are all examples of people who very well may know about jobs or opportunities that would suit you. Start asking several months before you hope to start a job or internship, because it will take some time to do research and arrange interviews. You can also use social media in your search. LinkedIn, for example, includes lots of searchable information on local companies. Follow and interact with people on social media to get their attention. Just remember to act professionally and communicate with proper grammar, just as you would in person.

With the right attitude, you will learn from each experience and get better with each subsequent interview. That should be your overarching goal. Consider these tips and tricks when interviewing, whether it be for a job, internship, college admission, or something else entirely:[2]

- Practice interviewing with a friend or relative. Practicing will help calm your nerves and make you feel more prepared. Ask for specific feedback from your friends. Do you need to speak louder? Are you making enough eye contact? Are you actively listening when the other person is speaking?
- Learn as much as you can about the company, school, or organization. Also be sure to understand the position for which you're applying. This will show the interviewer that you are motivated and interested in their organization.
- Speak up during the interview. Convey to the interviewer important points about you. Don't be afraid to ask questions. Try to remember the interviewer's name and call them by name.
- Arrive early and dress professionally and appropriately (you can read more about proper dress later in this chapter).
- Take some time to prepare answers to commonly asked questions. Be ready to describe your career or educational goals to the interviewer.

Common questions you may be asked during a job interview include these:

- Tell me about yourself.
- What are your greatest strengths?
- What are your weaknesses?
- Tell me something about yourself that's not on your resume.
- What are your career goals?
- How do you handle failure? Are you willing to fail?
- How do you handle stress and pressure?
- What are you passionate about?
- Why do you want to work for us?

Bring a notebook and a pen to the interview. That way you can take some notes, and they'll give you something to do with your hands.

Common questions you may be asked during a college admissions interview include these:

- Tell me about yourself.
- Why are you interested in going to college?
- Why do you want to major in this subject?
- What are your academic strengths?
- What are your academic weaknesses? How have you addressed them?
- What will you contribute to this college/school/university?
- Where do you see yourself in ten years?
- How do you handle failure? Are you willing to fail?
- How do you handle stress and pressure?
- Whom do you most admire?
- What is your favorite book?
- What do you do for fun?
- Why are you interested in this college/school/university?

Jot down notes about your answers to these questions, but don't try to memorize the answers. You don't want to come off too rehearsed during the interview. Remember to be as specific and detailed as possible when answering these questions. Your goal is to set yourself apart in some way from the other people they will interview. Always accentuate the positive, even when you're asked about something you did not like, or about failure or stress. Most importantly, though, be yourself.

Active listening is the process of fully concentrating on what is being said, understanding it, and providing nonverbal cues and responses to the person talking. It's the opposite of being distracted and thinking about something else when someone is talking. Active listening takes practice. You may find that your mind wanders and you need to bring it back to the person talking (and this could happen multiple times during one conversation). Practice this technique in regular conversations with friends and relatives. In addition to giving a better interview, it can cut down on nerves and make you more popular with friends and family, as everyone wants to feel that they are really being heard. For more on active listening, check out https://www.mindtools.com/CommSkll/ActiveListening.htm.

You should also be ready to ask questions of your interviewer. In a practical sense, there should be some questions that you have that you can't find the answer to on the website or in the literature. Also, asking questions shows that you are interested and have done your homework. Avoid asking questions about salary/scholarships or special benefits at this stage, and don't about anything negative that you've heard about the company or school. Keep the questions positive and relative to you and the position to which you're applying. Some example questions to potential employers include:

- What is a typical career path for a person in this position?
- How would you describe the ideal candidate for this position?
- How is the department organized?
- What kinds of responsibilities come with this job? (Don't ask this if they've already addressed this question in the job description or discussion.)
- What can I do as a follow-up?
- When do you expect to reach a decision?

See the sidebar in chapter 3 titled "Make the Most of School Visits" for some good example questions to ask the college admissions office. The important thing is to write your own questions related to answers you really want to know. This will show genuine interest. Be sure your question isn't answered on the website, in the job description, or in the literature.

EFFECTIVELY HANDLING STRESS

As you're forging ahead with your life plans, whether it's college, a full-time job, or even a gap year, you may find that these decisions feel very important and heavy and that the stress is difficult to deal with. First off, that's completely normal. Try these simple stress-relieving techniques:

- Take deep breaths in and out. Try this for thirty seconds. You'll be amazed at how it can help.
- Close your eyes and clear your mind.
- Go scream at the passing subway car. Or lock yourself in a closet and scream. Or scream into a pillow. For some people, this can really help.

- Keep the issue in perspective. Any decision you make now can be changed if it doesn't work out.

Want ways to avoid stress altogether? They are surprisingly simple. Of course, simple doesn't always mean easy, but it means they are basic and make sense with what we know about the human body:

- Get enough sleep.
- Eat healthy.
- Get exercise.
- Go outside.
- Schedule downtime.
- Connect with friends and family.

The bottom line is that you need to take time for self-care. There will always be conflict, but how you deal with it makes all the difference. This only becomes increasingly important as you enter college or the workforce and maybe have a family. Developing good, consistent habits related to self-care now will serve you all your life!

NADEAN TANNER:
CYBERSECURITY TRAINER AND INSTRUCTOR

Nadean Tanner has been in the technology industry for more than twenty years in a variety of positions, from marketing to training to web development to hardware. She has worked in academia as an IT director of a private elementary/middle school and as a technology instructor teaching postgraduate classes at the university level. She has trained in the corporate world for Fortune 50 companies as well as gotten her hands-on experience working for the Department of Defense with

Nadean Tanner

a focus on advanced cybersecurity. She is currently the senior manager of Global Technical Education Programs at Puppet in Portland, Oregon.

Can you explain how you became interested in information security as a career path?

It chose me really. It's been a journey of what looked interesting and challenging. As a child, when I got bored, I got into trouble. I love it because it's ever evolving. There's no time to get bored. I started in infrastructure, OS, and hardware and then it turned into security. Cybersecurity was so interesting and engaging, and trying to stay one step ahead of the bad guys was challenging. I started by teaching A+ and Met+ and then they needed someone who could teach CISSP (Certified Information Systems Security Professional certification), so I had a month to learn so I could teach the class. So I did! My career has bounced between doing and teaching all over the world. I usually end up back in the classroom teaching what I know.

What's a "typical" day in your job?

I have moved into project management and product development. I currently have a position at Puppet, which is more an automation, programming product that automates many security processes. I was teaching before that and was hearing from so many customers—what's next, what's next, what's next? I told them that one day, we'd have a tool that would find and fix the bad things automatically and that's what Puppet does.

My roles include evaluating curriculum, visiting customers, looking at how they are using our tool, making sure that we focus on lessons learned, seeing what went well, what can we learn, and how can we teach that to the next customer.

You have to be forward thinking. In cybersecurity, you have to keep one step ahead. We have to right all the time, whereas the bad guy has to be right only once. We have to be "on" all the time. They want to find weaknesses and holes. Hackers target end users especially. People are the weakest link. In fact, when I was at a hacking conference, I got a call about fraudulent credit card purchases. Somewhere I had used that card had been compromised.

Is the job what you expected it would be?

It's more exciting than I even thought. If you like to do puzzles and be strategic and really examine how someone might take advantage of a system, it's for you! It's very intriguing and exciting to be able to learn new things. My area of computers is interesting and exciting. I really enjoy it.

What's the best part of being in this field?

When I find people who are part of my tribe and understand the nuances of what I am talking about, whether at a conference or when teaching a class. Anywhere in the world you can find people that have the passion that you have.

What's the most challenging part of your job?

Remembering everything! Sometimes you get everything right, but catastrophe still ensues. But you just need to learn from those mistakes. We walk away smarter and stronger.

Do you think the current education adequately prepares students to enter this field?

You can get the math and learning experiences. It's easy to get a theoretical education and learn how it all "should" work, but that isn't always how it ends up. But until you've deciphered hexadecimal code yourself or actually configured a firewall, for example, you really don't know. Hands-on experience is critical, because even certifications are theoretical. You have to get your hands dirty to experience it and *then* that's when you really learn. The best of the best really want to share what they know.

What are some things in this profession that are especially challenging right now? What factors are affecting this job right now?

The lack of control over other people's computers due to IoT (Internet of Things) and the cloud. There's a huge gap between people who have great ideas and how these ideas end up. This is an advanced and consistent threat. Some of this software is extremely easy to take advantage of. Biomedical devices have been proven to be at risk, and self-driving cars can be hacked and can kill people.

IoT is not very secure and it's everywhere. Many programmers are creating apps and don't even know about basic hacking. They are creating an app and don't even know about exploits. IoT has so many different languages, and programmers aren't trained to care about security. It's more about the almighty dollar—get it out on the market.

Hardwired networks are generally more secure than Wi-Fi, but it really depends on the encryption and who sets it up. The cloud is the same—it all depends on who manages the security. A cloud is just someone else's computer. Microsoft, Google, etc.—even the big guys mess up. You can't be sure that things are secure, and the hackers are relentless.

Where do you see the field going in the future?

There will be more and more automated security and AI. I think true AI is around the corner. I see it becoming a bigger and greater part of our lives. I see it growing and compounding. It's not slowing down anytime soon.

What traits make for a good security analyst?

You need to have curiosity, learn from your mistakes, and be open to learning new things. You are done in this industry when you think you know it all. As the saying goes, "If you are the smartest person in the room, you are in the wrong room." Learn from others and hone your skill set. There are many types of cyber and you have to hone your skills. You also *have* to be passionate about it.

What advice do you have for young people considering this career?

Get a strong foundation, don't get intimidated, and learn from every vehicle you can—online, etc. Dabble and experiment and *then* decide what you don't like. You will find something that suits your aptitude and find your niche. You need to find a mentor too—they can guide you and tell you what to try. Join as many nonprofit cybersecurity organizations that you can. Security BSides, for example, holds conferences that are very inexpensive and you get a day's learning from very knowledgeable people in the field. Also, go to the library. Start with Python if you want to learn a language.

Mentoring is great. They can spark the interest in you and bring you along. Networking is huge!

How can a young person prepare for this career while in high school?

Of course, take STEM classes (math and science especially). Do the foundational education. Get a viewpoint and learn how things can change. History is great! You have to know where we came from. We have so many IT people with extreme specialties that don't translate elsewhere. You learn about mistakes others made that you don't have to make. Get an extremely well-rounded education. Jump in with both feet! It will come.

Any closing comments?

We are in one of the most viable industries! It's ever evolving, but job security is *excellent*. I feel like I could get a job anywhere in no time. It's a way of life. We have to continually protect what hackers are trying to take from us.

Dressing Appropriately

It's important to determine what is actually appropriate in the setting of the interview. What is appropriate in a corporate setting may be different from what you'd expect at a small liberal arts college or at a large hospital setting. Most college admissions offices suggest "business casual" dress, for example, but depending on the job interview, you may want to step it up from there.

Do the proper research to find out exactly how you should dress for your interview. *Getty Images/ PeopleImages*

Again, it's important to do your homework and come prepared. In addition to reading up on their guidelines, it never hurts to take a look around the site if you can to see what other people are wearing to work or to interviews. Regardless of the setting, make sure your clothes are not wrinkled, untidy, or stained. Avoid flashy clothing of any kind.

The term "business casual" means less formal than business attire like a suit, but a step up from jeans, T-shirt, and sneakers:

- *For men:* You can't go wrong with khaki pants, a polo or button-up shirt, and brown or black shoes.
- *For women:* Nice slacks, a shirt or blouse that isn't too revealing, and nice flats or shoes with a heel that's not too high

Follow-Up Communication

Be sure to follow up, whether in email or via regular mail, with a thank-you note to the interviewer. This is true whether you're interviewing for a job or an internship, or interviewing with a college. A handwritten thank-you note, posted in the actual mail, is best. In addition to being considerate, it will trigger the interviewer's memory about you and it shows that you have genuine interest in the position, company, or school. Be sure to follow the business-letter format and highlight the key points of your interview and experience at the company/university. Be prompt with your thank-you! Put it in the mail the day after your interview (or send that email the same day).

What Employers Expect

Regardless of the job, profession, or field, there are universal characteristics that all employers (and schools, for that matter) look for in potential employees. At this early stage in your professional life, you have an opportunity to recognize which of these foundational characteristics are your strengths (and therefore highlight them in an interview) and which are weaknesses (and therefore continue to work on them and build them up).

> Always aim to make your boss's job easier, not harder. Keeping this simple concept in mind can take you a very long way in the business world. By the same token, being able to convince an employer that you love to learn new things is one of the best ways to turn yourself into a candidate they won't be able to pass up.

Consider these universal characteristics that all employers look for:

- Positive attitude
- Dependability
- Desire to continue to learn
- Initiative
- Curiosity
- Effective communication
- Cooperation
- Organization

This is not an exhaustive list, and other characteristics can very well include things like being sensitive to others, being honest, having good judgment, being loyal, being responsible, and being on time. Specifically in infosec, you can add self-motivation, patience, perseverance, attention to detail, and self-control. Consider these important characteristics when you answer the common questions that employers ask. It pays to work these traits into the answers, of course being honest and realistic about yourself.

BEWARE WHAT YOU SHARE ON SOCIAL MEDIA

Most of us engage in social media. Sites such as Facebook, Twitter, Snapchat, and Instagram provide us a platform for sharing photos and memories, opinions, and life events, and reveal everything from our political stance to our sense of humor. It's a great way to connect with people around the world, but once you post something, it's accessible to anyone—including potential employers—unless you take mindful precaution.

Your posts may be public, which means you may be making the wrong impression without realizing it. More and more, people are using search engines like Google to get a sense of potential employers, colleagues, or employees, and the impression you make online can have a strong impact on how you are perceived. According to CareerBuilder.com, 60 percent of employers search for information on candidates on social media sites.3

Glassdoor.com offers the following tips for how to avoid your social media activity from sabotaging your career success:[4]

- Check your privacy settings. Ensure that your photos and posts are only accessible to the friends or contacts you want to see them. You want to come across as professional and reliable.
- Rather than avoid social media while searching for a job, use it to your advantage. Give future employees a sense of your professional interest by "liking" pages or joining groups of professional organizations related to your career goals.
- Grammar counts. Be attentive to the quality of writing of all your posts and comments.
- Be consistent. With each social media outlet, there is a different focus and tone of what you are communicating. LinkedIn is very professional while Facebook is far more social and relaxed. It's okay to take a different tone on various social media sites, but be sure you aren't blatantly contradicting yourself.
- Choose your username carefully. Remember, social media may be the first impression anyone has of you in the professional realm.

The best way to figure out what you want to do for a living is to get out there and start trying things! *Getty Images/BrianAJackson*

Summary

Well, you made it to the end of this book! Hopefully, you have learned enough about the infosec field to start along your journey, or to continue with your path. If you've reached the end and you feel like cybersecurity is your passion, that's great news. Or, if you've figured out that it isn't the right field for you, that's good information to learn too. For many of us, figuring out what we *don't* want to do and what we *don't* like is an important step in finding the right career.

"If you want to join the infosec field, there will always be a job for you! We really need your help and we need and want more good people in the field. I hope this book helps you choose a career path in infosec!"—Tanya Janca, founder, security trainer, and coach of SheHacksPurple.dev, specializing in training others in software and cloud security

There is a lot of good news about the infosec field, and it's a very smart career choice for anyone with a passion for computers. It's a great career for people who get energy from solving problems. Job demand is very strong. No matter which field you choose in infosec, demand is very high for this career! Whether you decide to be a hard-core cryptographer or want to focus on software security, having a plan and an idea about your future can help guide your decisions. We hope that by reading this book, you are well on your way to having a plan for your future. Good luck to you as you move ahead!

Notes

Introduction

1. "What Is Information Security?" GeeksforGeeks, https://www.geeksforgeeks.org/what-is-information-security/.

2. "Cybersecurity Roles and Job Titles," Department of Computer Science, George Washington University, https://www.cs.seas.gwu.edu/cybersecurity-roles-and-job-titles.

3. "Information Security Analysts," *Occupational Outlook Handbook*, US Bureau of Labor Statistics, https://www.bls.gov/ooh/computer-and-information-technology/information-security-analysts.htm.

4. Ibid.

Chapter 1

1. "What Information Security Analysts Do," *Occupational Outlook Handbook*, US Bureau of Labor Statistics, https://www.bls.gov/ooh/computer-and-information-technology/information-security-analysts.htm#tab-2.

2. "Difference between a Security Analyst and a Security Engineer?," Stack Exchange, https://security.stackexchange.com/questions/140959/difference-between-a-security-analyst-and-a-security-engineer.

3. "What Does a Security Engineer Do?," Career Explorer, https://www.careerexplorer.com/careers/security-engineer/.

4. "What Does a Security Software Developer Do?," Career Explorer, https://www.careerexplorer.com/careers/security-software-developer/.

5. Monika Agrawal, and Pradeep Mishra, "A Comparative Survey on Symmetric Key Encryption Techniques," *International Journal on Computer Science and Engineering* 4, no. 5 (May 2012): 877–82.

6. "Career Profile: Cryptologist," Careersthatdontsuck.com, February 24, 2007, https://careersthatdontsuck.com/2007/02/24/career-profile-cryptologist/.

7. "Information Security Analysts," *Occupational Outlook Handbook*, US Bureau of Labor Statistics, https://www.bls.gov/ooh/computer-and-information-technology/information-security-analysts.htm.

8. Ibid.

Chapter 2

1. "Explore Cybersecurity Jobs," Cyber Degrees, last updated March 30, 2020, https://www.cyberdegrees.org/jobs/.

2. "Cyber Security Degrees & Careers: How to Work in Cyber Security," Learn How to Become, https://www.learnhowtobecome.org/computer-careers/cyber-security/.

3. Lou Adler, "Survey Reveals 85% of All Jobs Are Filled via Networking," LinkedIn.com, February 29, 2016, https://www.linkedin.com/pulse/new-survey-reveals-85-all-jobs-filled-via-networking-lou-adler/.

Chapter 3

1. "Gap Year Data and Benefits," Gap Year Association, https://www.gapyearassociation.org/data-benefits.php.

2. Peter Van Buskirk, "Finding a Good College Fit," *U.S. News & World Report*, June 13, 2011, https://www.usnews.com/education/blogs/the-college-admissions-insider/2011/06/13/finding-a-good-college-fit.

3. Allison Wignall, "Preference of the ACT or SAT by State (Infographic)," CollegeRaptor, November 14, 2019, https://www.collegeraptor.com/getting-in/articles/act-sat/preference-act-sat-state-infographic/.

4. "Average Net Price by Sector over Time," College Board, https://research.collegeboard.org/trends/college-pricing/figures-tables/average-net-price-sector-over-time.

5. "FAFSA Changes for 2017–2018," Federal Student Aid, An Office of the U.S. Department of Education, https://studentaid.ed.gov/sa/about/announcements/fafsa-changes.

Chapter 4

1. Joshua Waldman, *Job Searching with Social Media for Dummies*, 2nd ed. (Hoboken, NJ: Wiley, 2013), 149.

2. Justin Ross Muchnick, *Teens' Guide to College & Career Planning*, 12th ed. (Lawrenceville, NJ: Peterson's, 2015), 179–80.

3. CareerBuilder.com, press releases, http://www.careerbuilder.com/share/aboutus/pressreleasesdetail.aspx?ed=12%2F31%2F2016&id=pr945&sd=4%2F28%2F2016.

4. Alice E. M. Underwood, "9 Things to Avoid on Social Media While Looking for a New Job," Glassdoor.com, January 3, 2018, https://www.glassdoor.com/blog/things-to-avoid-on-social-media-job-search/.

Glossary

accreditation: The act of officially recognizing an organizational body, person, or educational facility as having a particular status or being qualified to perform a particular activity. For example, schools and colleges are accredited. *See also* certification.

ACT: The American College Test (ACT) is one of the standardized college entrance tests that anyone wanting to enter undergraduate studies in the United States should take. It measures knowledge and skills in mathematics, English, reading, and science reasoning, as they apply to college readiness. There are four multiple-choice sections. There is also an optional writing test. The total score of the ACT is 36. *See also* SAT.

algorithm: The series of steps/formula/process that is followed to arrive at a result.

antivirus software: Computer programs that prevent, detect, and remove malware/viruses.

application: A computer program written in a programming language that serves a coordinated purpose for its users, such as a word processing application. Other examples of applications include spreadsheet apps like Excel, web browsers that enable you to view web pages on the internet, and graphics editors like Photoshop. *See also* web app; code.

associate's degree: A degree awarded by community or junior colleges that typically requires two years of study.

authentication: Verifying that someone is who they say they are. There are several levels (factors) of authentication. Single factor, two factor, three factor, and so on.

authorization: The collection of rights, permissions, and privileges that are assigned to users. It is what users are authorized to do within a secured infrastructure.

bachelor's degree: An undergraduate degree awarded by colleges and universities that's typically a four-year course of study when pursued full-time. However, this can vary by degree earned and by the university awarding the degree.

backdoor: An opening left in a program or software application (usually by the developer) that allows additional access to data. Typically, these are created for debugging purposes and aren't documented. Before the product ships, the backdoors should be closed; when they aren't closed, security loopholes are the result. *See also* Trojan horse.

back-end: The back-end of a computer system typically refers to the portion that stores or manipulates data, and it is not usually visible or accessible to the users. Often also called the server side, this is the physical infrastructure or hardware. Back-end developers might write code for the server, the database, the operating system, or the actual software. *See also* front-end.

black box: A device whose internal circuits, makeup, and processing functions are unknown but whose outputs can be observed and analyzed. Black box penetration testing is done without any knowledge of how an organization is structured, what kinds of hardware and software it uses, or its security policies, processes, and procedures.

blacklist: A security approach that allows everything by default and denies access by exception.

bot: A malicious remote-control tool set up on systems to create botnets. Also called a zombie or agent. These tools can assimilate a work or personal computer into the botnet and add its resources (processing power, bandwidth, etc.) to the collective.

botnet: A massive deployment of malicious code onto numerous compromised systems that are all controlled by a hacker (often referred to as a bot herder).

brute force attack: A type of attack that relies purely on trial and error.

certification: The action or process of confirming certain skills or knowledge of a person. Usually provided by some third-party review, assessment, or educational body. Individuals, not organizations, are certified. *See also* accreditation.

chief information security officer (CISO): A high-level management position responsible for the entire information security (infosec) division/staff.

cipher: An algorithm (computer instructions) that helps someone encrypt sensitive information or messages so that unauthorized people can't read them, or that helps to decrypt messages that have been secured with an encryption method.

cloud computing: In a cloud computing system, the computer resources (such as storage and computing power) are available on demand at a centralized data center (called "the cloud"), not on the user's computer. Often, the internet serves as the cloud, but clouds can also be run from a centralized data center in a single organization (called an "enterprise" cloud) or there can be third-party clouds used by multiple organizations ("public" clouds). There can also be a combination of these two types, called "hybrid" clouds. Benefits of cloud computing include economies of scale (saving time or money by "going big"), reduced IT infrastructure costs, better response to changing demands, and quicker adaptations to updates and changes.

code: A set of instructions and commands read by a computer. Also called a programming language, code comes in many varieties for different uses, including assembly, compiled, interpreted, and object-oriented languages. *See also* programming language.

cookie: A plain-text file stored on your machine that contains information about you (and your preferences) for use by a database server. Cookies are frequently used for various legitimate purposes, but they can also be used by malicious websites to track user activities.

cryptographers/cryptologists/cryptanalysts: These professionals use encryption to secure information or to build security software, develop stronger encryption algorithms, and/or analyze the encrypted information to break the code/cipher or to determine the purpose of malicious software. *See also* digital forensics expert.

cybersecurity: A set of techniques put in place to protect the safety and integrity of networks and other internet-connected systems from attacks. This includes securing their hardware, software, and data.

data breach: The (usually unintentional) release of secure or private/confidential information to an untrusted environment.

data encryption program: Software that encrypts data in a file, network packet, or application so that it is secure and can't be accessed by unauthorized users.

database: A collection of data in a computer that's organized into tables and schemas and can be accessed and reviewed by using database queries and creating reports. The data is presented in a model that represents an actual physical object that's familiar to users, such as a folder. Commercially successful databases include Microsoft Access, SQL Server, and Oracle DB.

DDoS (distributed denial-of-service) attacks: Multiple machines work together to attack one target. DDoS attackers often use botnets to carry out large-scale attacks. *See also* botnet.

DevOps: Software development methods that attempt to shorten the time between making a change to a computer system and that change being incorporated into the system, all the while focusing on accuracy and quality. This philosophy emphasizes collaboration and communication between software developers (*Dev*) and other IT professionals (*Ops*). The goal is to deliver new features and updates, as well as fixes, quickly and often. DevOps uses the practices of continuous integration and continuous delivery, among other approaches, to achieve this mode of working. Similar to the Agile or Waterfall software development movements.

DevSecOps: The addition of security to a DevOps environment.

digital forensics expert: This professional investigates and analyzes digital information to help in investigations and solve computer-related crimes. These experts look into incidents of hacking, trace sources of computer attacks, and recover lost or stolen data.

doctorate degree: The highest level of degree awarded by colleges and universities. Qualifies the holder to teach at the university level. Requires (usually published) research in the field. Typically requires an additional three to five years of study after earning a bachelor's degree. Anyone with a doctorate degree can be addressed as a "doctor," not just medical doctors.

DoS (denial-of-service) attacks: Attacks that prevent all users—even legitimate ones—from using the system.

encryption: The process of encoding information so that only people who are approved to access it can do so. Unauthorized users cannot access the information without a key that decrypts the information. *See also* cipher; algorithm.

ethical hacker: This infosec professional legally breaks into computers and devices to test an organization's defenses, with the goal of making them more secure and resistant to nefarious hackers. They are also often called white hat hackers. *See also* penetration test.

exploit: A malicious application or script that can be used to take advantage of a computer's vulnerability. *See also* vulnerability.

firewall: The part of a computer system or network that attempts to block unauthorized access while still allowing acceptable communication. A firewall acts as a barrier between a trusted system or network and outside connections, such as the internet.

front-end: In context of IT, the front-end of a computer system usually refers to the interface layer at the very "front" that the user interacts with; also often called the client side or the presentation layer. Front-end development is typically concerned with how information is presented to the users. *See also* back-end.

full-stack developer: An IT engineer who can work with and manage all databases, servers, systems engineering components, and clients. Depending on the project, customers may need a mobile *stack*, a web *stack*, or a native application *stack*.

gap year: A gap year is a year between high school and college (or sometimes between college and postgraduate studies) whereby the student is not in school but is instead typically involved in volunteer programs, such as the Peace Corps, in travel experiences, or in work and teaching experiences.

grants: Money to pay for postsecondary education that is typically awarded to students who have financial needs, but can also be used in the areas of athletics, academics, demographics, veteran support, and special talents. Grants do not have to be paid back.

hacker: Person who tries to illegally break into a computer system for money, a social cause, fun, and so on. *See also* ethical hacker.

hardware: The physical parts of the computer system, such as the actual machines, cards, wires, keyboards, and other parts, both external and internal. If you can hold it in your hand, it's part of the hardware system. *See also* software; peripherals.

information security analysts: These IT professionals plan and carry out security measures to protect an organization's computer networks and systems.

malware: The collective name for several malicious software types, including viruses, ransomware, and spyware. It's shorthand for malicious software.

master's degree: A secondary degree awarded by colleges and universities that requires at least one additional year of study after obtaining a bachelor's degree. The degree holder shows mastery of a specific field. Teachers in public school settings are often required to pursue their master's degrees after having worked as a teacher for a prescribed amount of time.

network: A group of connected computers that exchange data with each other. They are connected via some kind of telecommunications network made up of wires, cables, or via wireless media. The internet is an example of a computer network.

network engineer: An IT professional who focuses on computer network and systems administration. Because administrators work with computer hardware and equipment, a degree in computer engineering or electrical engineering is often desirable in this field.

open source: A type of software in which the original source code is released under a license whereby the copyright holder grants users the rights to study, change, and distribute the software to anyone and for any purpose. Open source software may be developed in a collaborative public manner. Benefits of using open source software include lower software and hardware costs and solid support and maintenance from the open source community.

operating system (OS): This is the system software that controls the computer's hardware and software systems and defines how they interact. Examples of common operating systems include MacOS, Windows, Linux, and DOS.

Software applications that run on a specific computer are often OS dependent, in that they are written specifically for a certain OS and can't be run on any others. *See also* web app.

penetration test: An authorized, simulated attack on a computer system in order to evaluate the security of that system and find vulnerabilities in it for the benefit of the system owner. Cyberanalysts who perform these are called pen testers. *See also* ethical hacker.

peripherals: Specific types of computer hardware input or output devices that give computers additional functionality. Strictly speaking, they are not required for the computer to run. Examples include thumb drives, sound cards, speakers, printers, joysticks, mice, and yes, even keyboards. Peripherals cannot operate by themselves but need a computer in order to function. *See also* hardware.

personal statement: A written description of your accomplishments, outlook, interest, goals, and personality that's an important part of your college application. The personal statement should set you apart from others. The required length depends on the institution, but they generally range from one to two pages, or five hundred to one thousand words.

postsecondary degree: Educational degree above and beyond a high school education. This is a general description that includes trade certificates and certifications, associate degrees, bachelor's degrees, master's degrees, and beyond.

programmer: A general term for a person (or machine) who writes code and creates and tests computer programs. Programmers usually specialize in one or a few programming languages. Areas of expertise include software development, database development, hardware programming, and web development.

programming language: A formal language that communicates to a computer through its applications and programs. These programs are created from instructions and commands written in the programming language by developers. Examples of widely used programming languages include C#, Java, Python, and Visual Basic. *See also* code.

ransomware: A form of malware that prevents you from accessing files on your computer—holding that data hostage. It usually encrypts files and then requests that a ransom be paid in order to have them decrypted or recovered.

SAT: The Scholastic Aptitude Test (SAT) is one of the standardized tests in the United States that anyone applying to undergraduate studies should take. It measures verbal and mathematical reasoning abilities as they relate to predicting successful performance in college. It is intended to complement a student's GPA and school record in assessing readiness for college. The total score of the SAT is 1600. *See also* ACT.

scholarships: Merit-based aid used to pay for postsecondary education that does not have to be paid back. Scholarships are typically awarded based on academic excellence or some other special talent, such as music or art.

security administrators: They install and manage organization-wide security systems. They may also do some of the tasks of a security analyst in smaller organizations.

security analysts/engineers: These specialists analyze and evaluate weaknesses in the infrastructure (might be the software, hardware, networks, or cloud). They find the best tools and countermeasures to address those vulnerabilities. They may also assess the damage from security incidents and recommend solutions and best practices. They may also test for compliance with security policies and help the organization create, implement, or manage its security solutions.

security architects: They design the security system or major components of the security system, and they may head a security design team that's building a new security system.

security protocol: A protocol that performs a security-related function and applies cryptographic methods to secure data.

security software developers: They develop security software, including tools for monitoring, traffic analysis, intrusion detection, virus/spyware/malware detection, antivirus software, and so on. They may also integrate/implement security into applications and software.

software: The programs written in code/a programming language that run on the computer and operate its various functions. These are the instructions that tell the computer how to work and what to do and are usually platform/OS dependent. *See also* hardware; code.

spyware: Malware that accesses information about a person or organization without them knowing and uses that information to hack another system without the consumer's consent.

SQL injection: An attack that inserts or "injects" an SQL query from the client to the application for the purposes of reading sensitive data from the database or otherwise modifying database data for nefarious purposes.

support: The computer support professionals who provide advice and help to users having issues with their computer software and hardware. Their job is to troubleshoot, identify, and fix problems with single computers or networks under their purview.

Trojan horse: Malware that often allows a hacker to gain remote access to a computer through a backdoor. *See also* backdoor.

virus: A program made to damage a computer system. More sophisticated viruses are encrypted; they hide in a computer and may not appear until the user performs a certain action or until a certain date.

vulnerability: A system's/application's weakness, design problem, or implementation error that can lead to unexpected and undesirable events regarding security (such as unlawful access or data breach). *See also* exploit.

web app: An application that runs in a web browser. A common example is a webmail application like Gmail, which stores your account data (your email) in the Google cloud. You can access a web application from any computer connected to the internet using a standard browser. Web apps are typically platform/OS independent since the website serves as the user interface.

web development: A broad term that refers to the varied tasks involved in creating a website or web application, which will be hosted on the internet or on a local intranet. Web development includes designing the interface and the website; creating, programming, testing, and formatting the web content; client-side/server-side scripting for handling user interactions; managing and configuring network security; and more.

worm: Malware that replicates itself in order to spread its infection to other connected computers. It burrows into multiple systems like a worm.

Resources

*A*re you looking for more information about the professions in infosec, or even want to learn more about a particular area of it? Do you want to know more about the college application process or need some help finding the right educational fit for you? Do you want a quick way to search for a good college or school? Try these resources as a starting point on your journey toward finding a fulfilling career as an infosec specialist!

Books

Althoff, Cory. *The Self-Taught Programmer: The Definitive Guide to Programming Professionally.* San Francisco: Triangle Connection, 2017.

Bolles, Richard N. *What Color Is Your Parachute? 2019: A Practical Manual for Job Hunters and Career Changers.* Rev. ed. New York: Ten Speed, 2018.

Fiske, Edward. *Fiske Guide to Colleges.* Naperville, IL: Sourcebooks, 2018.

Guthrie, Julian. *Alpha Girls: The Women Upstarts Who Took On Silicon Valley's Male Culture and Made the Deals of a Lifetime.* New York: Penguin Random House, 2019.

Janca, Tanya. *Alice and Bob Learn Application Security.* Hoboken, NJ: Wiley, 2020.

Muchnick, Justin Ross. *Teens' Guide to College & Career Planning.* 12th ed. Lawrenceville, NJ: Peterson's, 2015.

Newnham, Danielle. *Female Innovators at Work: Women on Top of Tech.* London: Apress, 2016.

Pfleeger, Charles P., Shari Lawrence Pfleeger, and Jonathon Marguiles. *Security in Computing.* 5th ed. Upper Saddle River, NJ: Pearson Education, 2015.

Princeton Review. *The Best 382 Colleges, 2018 Edition: Everything You Need to Make the Right College Choice.* New York: Princeton Review, 2018.

Sikorski, Michael, and Andrew Honig. *Practical Malware Analysis: A Hands-On Guide to Dissecting Malicious Software.* San Francisco: No Starch, 2012.

Wheeler, Tarah. *Women in Tech: Take Your Career to the Next Level with Practical Advice and Inspiring Stories.* Seattle, WA: Sasquatch Books, 2017.

Websites/Blogs

American Gap Year Association

gapyearassociation.org

The American Gap Year Association's mission is "making transformative gap years an accessible option for all high school graduates." A gap year is a year taken between high school and college to travel, teach, work, volunteer, generally mature, and otherwise experience the world. The website has lots of advice and resources for anyone considering taking a gap year.

The Balance Website

www.thebalance.com

This site is all about managing money and finances, but also has a large section called "Your Career," which provides advice for writing resumes and cover letters, interviewing, and more. Search the site for "teens" and you can find teen-specific advice and tips.

Codecademy

www.codecademy.com

This website is an effective and easy way to learn to code. It includes short modules on HTML, CSS, and website development, after which you can move on to other programming languages. The courses are easy to follow, and they award badges when you finish each one, which can be a nice motivator to keep learning.

Codewars

www.codewars.com

This site is run by a community of developers who attempt to improve their development skills through challenges set by fellow members. It's similar to a coding dojo, where developers train with each other and help each other get better through practice. This is a great site once you have a little experience under your belt with beginner sites.

College Board Website

www.collegeboard.org

The College Board tracks and summarizes financial data from colleges and universities all over the United States. This site can be your one-stop shop for all things college research. It contains lots of advice and information about taking and doing well on the SAT and ACT tests, many articles on college planning, a robust college-searching feature, a scholarship-searching feature, and a major- and career-search area. You can type your career of interest (e.g., occupational therapy) into the search box and get back a full page that describes the career, gives advice on how to prepare, where to get experience, how to pay for it, what characteristics you should have to excel in this career, lists of helpful classes to take while in high school, and lots of links for more information. A great, well-organized site.

College Grad Career Profile Website

www.collegegrad.com/careers

Although this site is primarily geared toward college graduates, the career profiles area, indicated above, has a list of links to nearly every career you could ever think of. A single click takes you to a very detailed, helpful section that describes the job in detail, explains the educational requirements, includes links to good colleges that offer this career, includes links to actual open jobs and internships, describes the licensing requirements (if any), lists salaries, and much more.

GoCollege

www.gocollege.com

Calls itself the number one college-bound website on the internet. Includes lots of good tips and information about getting money and scholarships for college and getting the most out of your college education. Has a good section on how scholarships in general work.

Go Overseas

www.gooverseas.com

Claims to be your guide to more than fourteen thousand study and teach abroad programs that will change how you see the world. Also includes information about high school abroad programs, and gap year opportunities. Includes community reviews and information about finding programs specific to your interests and grade-level teaching aspirations. For example, it includes information about the best countries for teaching English abroad.

Kahn Academy
www.khanacademy.org
The Kahn Academy website is an impressive collection of articles, courses, and videos about many educational topics in math, science, and the humanities. You can search any topic or subject (by subject matter and grade), and read lessons, take courses, and watch videos to learn all about it. Includes test prep information for the SAT, ACT, AP, GMAT, and other standardized tests. There is also a "College Admissions" tab with lots of good articles and information, provided in the approachable Kahn style.

Krebs on Security
krebsonsecurity.com
Brian Krebs worked as a reporter for the *Washington Post* from 1995 to 2009, authoring more than thirteen hundred blog posts for the *Security Fix* blog. This is his personal blog, which covers in-depth security news and investigations. The focus is on cybercrime and other major data breaches and hacks.

Live Career Website
www.livecareer.com
This site has an impressive number of resources directed toward teens for writing resumes, writing cover letters, and interviewing.

Mapping Your Future
www.mappingyourfuture.org
This site helps young people figure out what they want to do and maps out how to reach career goals. Includes helpful tips on resume writing, job hunting, job interviewing, and more.

#MentoringMonday on Twitter
Pairs less experienced people with professional mentors. Many, many infosec people are participating. They message each other to mentor each other, answer questions, help find employment, and more. It includes all areas of technology,

especially infosec. It's a grassroots movement and it's all free. Use the hashtag #MentoringMonday to post questions and find a mentor.

Modern Analyst

www.modernanalyst.com

Modern Analyst is a community and resource portal for business analyst and systems analyst professionals. This site includes webinars, forums, job postings, and more.

Monster

www.monster.com

Perhaps the most well-known and certainly one of the largest employment websites in the United States. You fill in a couple of search boxes and away you go. You can sort by job title, of course, as well as by company name, location, salary range, experience range, and much more. The site also includes information about career fairs, advice on resumes and interviewing, and more.

Network World

www.networkworld.com

This magazine-style website focuses on network news, trend analysis, and product testing. It also includes many of the industry's most important blogs. It's considered one of the most active and frequent publishers of network-related content on the internet.

Occupational Outlook Handbook

www.bls.gov/ooh

The US Bureau of Labor Statistics produces this website. It offers lots of relevant and updated information about various careers, including average salaries, how to work in the industry, the job's outlook in the job market, typical work environments, and what workers do on the job. See www.bls.gov/emp/ for a full list of employment projections.

Open Web Application Security Project

owasp.org

An online community since 2001 that produces freely available articles, methodologies, documentation, tools, and technologies in the field of web application security. They hold conferences and workshops around the world.

Peterson's College Prep Website
www.petersons.com
In addition to lots of information about preparing for the ACT and SAT tests and easily searchable information about scholarships nationwide, Peterson's site includes a comprehensive searching feature to search for universities and schools based on location, major, name, and more.

Princeton Review Website
www.princetonreview.com/quiz/career-quiz
This site includes a very good aptitude test geared toward high schoolers to help them determine their interests and find professions that complement those interests.

SheHacksPurple.dev
A learning platform created by industry leader Tanya Janca dedicated to teaching application security, DevSecOps, and cloud security. Tanya creates all the content and performs all in-person and online training and coaching. Sign up for regular online content updates and courses at reasonable (and sometimes free) prices.

Study.com Website
www.study.com
A site similar to Kahn Academy where you can search any topic or subject and read lessons, take courses, and watch videos to learn all about it.

TeenLife: College Preparation
www.teenlife.com
This organization calls itself "the leading source for college preparation," and it includes lots of information about summer programs, gap year programs, community service, and more. They believe that spending time out "in the world" outside of the classroom can help students develop important life skills. This site contains lots of links to volunteer and summer programs.

U.S. News & World Report *College Rankings*
www.usnews.com/best-colleges
U.S. News & World Report provides almost fifty different types of numerical rankings and lists of colleges throughout the United States to help students

with their college search. You can search colleges by best reviewed, best value for the money, best liberal arts schools, best schools for B students, and more.

W3Schools
www.w3schools.com
This well-organized site is easy to navigate and contains tutorials on all topics programming related. It includes lots of examples as well. Topics include learn HTML, learn CSS, learn Bootstrap, learn JavaScript, learn C++, learn XML, and many, many more. This site is frequented by newbies and senior developers alike, due to its depth and breadth of topics.

WoSEC: Women of Security
https://wearetechwomen.com/wosec-women-of-security/
Founded in founded in 2015 by Tanya Janca to help women working in technology maximize their potential. One of their goals is to contribute to an increase in women working in the technology industry (which is currently 15 percent). They host events and conferences and have a membership of more than seventeen thousand diverse women working across a multitude of industries and tech disciplines.

Podcasts

The Big Tech Show
Every week they analyze the big tech issues, interview industry leaders, and review the hottest gadgets. Hosted by Adrian Weckler.

The CyberJungle
Cohosted by digital forensic analyst Ira Victor and Samantha Stone, an award-winning journalist who also produces the show. Ira Victor is a sought-after expert in the cybersecurity realm, and Samantha Stone is a veteran broadcaster and reporter who specializes in politics and legislation. They provide an entertaining and informative take on the latest in security news and talk with thought leaders and insiders who can weigh in on current happenings.

Greater Than Code
For a long time, tech culture has focused too narrowly on technical skills; this has resulted in a tech community that too often puts companies and code over people. *Greater Than Code* is a podcast that invites the voices of people who are not heard from enough in tech—women, people of color, and trans and/ or queer folks—to talk about the human side of software development and technology.

How I Built This, with Guy Raz
In this NPR app, Guy Raz dives into the stories behind some of the world's best-known companies. *How I Built This* weaves a narrative journey about innovators, entrepreneurs, and idealists—and the movements they built.

Risky Business
Established in 2007, *Risky Business* is one of the most highly regarded and most-listened-to podcasts in the information security space. *Risky Business* aims to take a lighthearted look at information security news and has earned a reputation for covering the most alarming hacks and gaining insights from guests in the know.

Security Now!
A weekly podcast featuring Steve Gibson and Leo Laporte, who spend nearly two hours discussing vital security concerns ranging from news to long-standing issues, concerns, and solutions. *Security Now!* focuses on personal security, offering valuable insights to help their audience overcome common challenges and ramp up their personal security.

Security Weekly
Covers IT security news, vulnerabilities, hacking, research, and related topics of interest for the IT community by serving as a security podcast network. Their goal is to reach a wide global audience to share information and insights that help them learn, grow, and become savvy IT professionals. Hosted by Paul Asadoorian, various cohosts, and special guests, *Security Weekly* has been going strong for eleven years.

Tech Talks
This is a series of podcasts discussing the latest tech news and trends, and featuring interviews with tech leaders who share their experiences.

Women Tech Charge
Hosted by Dr. Anne-Marie Imafidon, the CEO of stemettes.org, the series looks at the incredible women revolutionizing our lives through technology.

About the Author

Kezia Endsley is a writer and editor from Indianapolis, Indiana. In addition to editing technical publications and writing books for teens, she enjoys running and triathlons, traveling, reading, and spending time with her family and many pets.